WITHDRAWN FROM
TSC LIBRARY

UP FOR GRABS

UP FOR GRABS

The Future of the Internet I

LEE RAINIE,
JANNA QUITNEY ANDERSON,
AND
SUSANNAH FOX

CAMBRIA
PRESS
AMHERST, NEW YORK

Copyright 2008 Lee Rainie, Janna Quitney Anderson, and Susannah Fox

All rights reserved
Printed in the United States of America

No part of this publication may be reproduced, stored in or introduced into a retrieval system, or transmitted, in any form, or by any means (electronic, mechanical, photocopying, recording, or otherwise), without the prior permission of the publisher.

Requests for permission should be directed to:
permissions@cambriapress.com, or mailed to:
Cambria Press
20 Northpointe Parkway, Suite 188
Amherst, NY 14228

Date of Initial Release: January 9, 2005.

Library of Congress Cataloging-in-Publication Data

Rainie, Harrison.
   Up for grabs : the future of the Internet I / Lee Rainie, Janna Quitney Anderson, and Susannah Fox.
      p. cm.
   Includes bibliographical references and index.
   ISBN 978-1-60497-517-8 (alk. paper)
   1. Internet—Social   aspects. 2. Internet—Forecasting. 3. Internet—History. I. Anderson, Janna Quitney. II. Fox, Susannah. III. Title. IV. Title: Future of the Internet I.

  HM851.R36 2008
  302.23'1—dc22

2008005644

For Ithiel de Sola Pool

Pool, a leading communications researcher of the 20th century, died in the mid-1980s, but he was a significant influence when vital decisions were being made regarding freedom on the Internet in the decades to follow. He inspired many Internet stakeholders with his book, *Technologies of Freedom*. In it, he projected that interconnected computing devices would be joined to form an open-ended, all-encompassing structure. He described it as "the largest machine that man has ever constructed—the global telecommunications network; the full map of it no one knows; it changes every day," and he projected that it would be questioned by regulators fearing the challenge to the economic and political status quo. He warned that a positive future would be delayed if regulators chose to interfere. His 1983 book, *Forecasting the Telephone*, was an inspiration for the initiation of the "Imagining the Internet" project and surveys.

Table of Contents

List of Tables	ix
Foreword	xi
Acknowledgments	xvii
Summary of Findings	xix
Part 1: Introduction	1
Part 2: Institutions	5
Part 3: Social Networks	35
Part 4: Network Infrastructure	53
Part 5: Digital Products	69
Part 6: Civic Engagement	83
Part 7: Embedded Networks	103
Part 8: Formal Education	119
Part 9: Democratic Processes	137
Part 10: Families	149
Part 11: Extreme Communities	163
Part 12: Politics	173

Table of Contents

Part 13: Health System Change — 187

Part 14: Personal Entertainment — 203

Part 15: Creativity — 217

Part 16: Internet Connections — 231

Part 17: Looking Back; Looking Forward — 241

Part 18: Reflections — 295

Methodology — 299

Appendix I: Future Scenarios Inspired by Early 1990s "Awe Stage" Projections — 305

Appendix II: Biographies — 351

Appendix III: Survey Results and Methodology — 377

Contributor Index — 391

Subject Index — 395

List of Tables

Table 1. Overall survey completion rate. 301

Table 2. Overall survey completion rate. 389

Foreword

When Janna Quitney Anderson, Susannah Fox, and I began to think about surveying technology experts about the future, we worried there might be too much consensus among the *digerati*. We fretted there was a firm conventional wisdom among the tech community's tribal leaders about the future of the Internet and what its social, political, and economic impacts would be. Our concern was that even the most provocative questions would yield answers that would be too uniform and too predictable to be interesting and useful to anyone.

That was a lot of wasted worry.

The experts' answers for this volume suggest they more or less agree about where digital hardware and software are heading, and there is little dispute that these technologies will become more important in users' everyday lives. Beyond that, though, there is a rich discussion and often raucous disagreement about how the new, improved technology will be used, how people and institutions will respond to tech changes, and the degree to which the impact will be beneficial or harmful.

Thus, an overview of key findings shows that the respondents to our survey believe:

- There is a strong likelihood of an attack on the Internet or underlying power system in the coming decade that could have serious economic consequences, though there is much disagreement about whether any great number of people would be in physical jeopardy as a result of that attack.
- An "Internet of things" will emerge as inanimate items and the environment itself will become connected in the network, and more surveillance of people by governments and businesses will result. Still, there is a significant dispute about how much surveillance democratic societies will permit and whether these changes will even generate much protest.
- The processes of formal education will change as information in all forms becomes easily available to people, though there is a large cry of despair that, to date, K–12 schools and even universities have not adjusted to this new reality so far. Indeed, these experts believe that educational institutions rank high with health institutions as the least receptive to the changed reality offered by the digital era.
- Individual creativity will flourish as cheaper tools become available to produce and distribute "knowledge goods" and media. Yet the capacity of these neo-creators to disseminate their ideas or have them take hold with any kind of audience is uncertain as the volume of digital material explodes and users adjust to information overload by aggressively filtering the material they encounter. Further, there is notable uncertainty about whether markets dominated by major commercial enterprises or governments will allow much sharing of digital material.
- Social groups of all kinds—constructive and destructive alike—have new ways to recruit and interact, though there is no agreement on whether this is changing the fundamental contours of social engagement or the amount of meaningful

social activity. These respondents are relatively confident that there will be more buzz in the social sphere, but they do not clearly come down on one side or another about whether this new chatter is deepening the pool of social capital.

If there is such contention among experts about the coming impact of the Internet, why consult them? For starters, many of the people who responded to this survey are the builders of the Internet. They have the capacity to act on what they believe or influence those who set policy. Alone, their voices matter, no matter how disputatious they are.

Second, this effort resulted in a unique kind of collective intelligence. No one knows everything. Everyone knows something. Together, they know an amazing amount—and in the past, many of the respondents have been right more than they have been wrong. Indeed, our inspiration for the predictions survey came after we had studied experts' predictions from the early 1990s about the Internet. As a group, they were so prescient and provocative in their statements that it was reasonable to believe their opinions about the world of 2014 would be revealing and helpful.

Third, like a good time capsule, this volume contains artifacts that can be fun to examine and important to study. Or, if you prefer another metaphor, this is a thorough audit of the state of thinking about the Internet in the early 21st century. The issues hashed out here are those that should be raised and discussed in computer science programs, public policy forums, sociology and communications classes, and business school case studies.

Fourth, and arguably most important, these predictions can be touchstones for those trying to live in the digital age and for those researching it. People who are anxious to build civil society, to sustain thriving enterprises, to promote their ideas, their causes, and their culture will find innumerable insights here that can help them see how to position themselves in the coming years. Researchers have a lode of ideas to assess in their own studies, to push back against, or to undergird their own notions about the world to come. This volume can be a rich reference source to those who are pondering issues related to the digital divides, the world of intellectual property, the effect of Internet use on a wide range of organizations and social structures, the first decade of consumer use of the Internet, and the deployment of technology in years to come. Futurology can be a fun parlor game in many hands. In this work, it is a serious endeavor at creating a baseline of predictions that can truly be assessed as the Internet evolves.

The researchers who use this as a blueprint for their own work would be wise to embrace a recurring notion that is often explicit in these predictions: The very thing they are dissecting will change. The Internet of the future will not function on the same architecture, with the same protocols and same gear that makes the Internet function now. There is a strong sense among these respondents that the thing they are being asked to assess will be improved to address concerns that its creators did not fully anticipate when they were building these technologies in the 1980s and early 1990s. Neither the capacity of clever people with bad intentions to exploit the Internet nor the need for a network system that would efficiently cope with ever-greater loads of digital material were considered at the early design stages. These experts sense that a new Internet should be constructed and yet they are not fully sure what its capacities will be. And a striking diffidence runs through some of their comments as they suggest that some basic "givens" of the future are not known now.

They also express concerns about the policy uncertainties that will determine the course of the future once they are resolved. How will intellectual property be handled in an era where increasing numbers of users are creating their own digital goods, publishing their ideas, mashing up material they have found from other sources online, and sharing it through peer-to-peer systems? What is theft and what is "fair use"? How can the legal system stay abreast of the claims of ownership that multiply daily? At an even more fundamental level, there are questions about who will control the flows of material that traverse the Internet.

Not only do they express uncertainties about the shape of the Internet's infrastructure and the way digital property is handled, these respondents are also awestruck by changes that developers and markets have introduced that they failed to anticipate. It starts with the creation of the World Wide Web itself and how this layer of activity on the Internet has been stocked with digital material beyond their imaginings. Few foresaw the rise of peer-to-peer services and how they would be built initially on the sharing of digital music files. Some expected e-commerce to develop, but the speed of its ascension was a cause of wonder to some of these respondents. As expert as they are, they are capable of being charmed by surprise.

By the same token, some of them were humbled to note that the things they thought would happen did not occur—or at least occur at the pace they thought. Some felt they were misled by their own hubris as they thought that all social and economic institutions would bend to their creation. In fact, they noted that neither educational nor medical organizations had responded quickly to either the efficiencies offered by digital technology or the way it was supposed to empower individuals as they encountered institutions or functioned inside them.

There is a danger in an exercise like this that technological determinism will permeate the findings. This work was launched, after all, by the Pew Internet & American Life Project, an organization whose premise is that the Internet, as a tool, possibly influences human endeavors and that there is a need to produce research testing the the power and degree of those influences. Moreover, the questions and predictions being assessed in the survey here were very much focused on the direction of technology change and how that might affect the social universe. Nevertheless, respondents took every opportunity to assert many versions of a single theme: Human determinism would be the dominant force in the future.

That idea takes several forms. Mack Reed of the Digital Government Research Center at USC spoke for many when he invoked the grand humanist notion that there are "timeless constants" guiding human activities. In many of their answers, these experts embrace the thought that humans have innate limits and have a need to set limits. People can only reasonably manage so many relationships. They have a fixed amount of time to devote to their work and play. They have limited energy to act on the things that matter to them in their personal worlds and the wider public world beyond their doors. Their attention spans and capacity to process information are finite. To be sure, these respondents are pretty sure that some of the boundaries on human energy and capacity will be adjusted. Yet they are equally sure people are not going to be fundamentally rewired or reoriented in this new era even as "data electricity" is coursing all around them.

A similar perception about human agency emerges from the discussion here about the politics of technology. These respondents say in a number of ways that organizations can only handle a certain amount of change. In this group, there is fascination—and a decent amount of concern—about the role that major corporations have played in the build-out of the Internet and their intentions for using the Internet in the future.

And yet…there are voices here that dispute that thought. That is the exciting part of conducting these surveys and being in conversation with so many enthusiasts about the future of the Internet. This book does not make a grand argument. It reflects, instead, the "culture by commotion" that is central to these times as disruptive technologies work their way through people's lives and their organizations. In the spirit of the age of interactive technology, this book is also meant to provoke your reaction and inspire your own contribution. We invite you to add your own insights to the collective intelligence amassed at http://www.imaginingtheinternet.org.

Lee Rainie, Director
Pew Internet & American Life Project

ACKNOWLEDGMENTS

PEW INTERNET & AMERICAN LIFE PROJECT

The Project is a nonprofit, nonpartisan "fact tank" that explores the impact of the Internet on children, families, communities, the workplace, schools, health care, and civic/political life. The Project takes no positions on policy questions. Support for the Project is provided by The Pew Charitable Trusts and it is an initiative of the Pew Research Center. The Project's Web site is http://www.pewinternet.org.

At The Pew Charitable Trusts, the support of two people in particular has sustained the Project: President and CEO Rebecca Rimel and the Managing Director for Information and Civic Initiatives Donald Kimelman. Inside the Pew Research Center, the Project profits handsomely from the expertise of President Andrew Kohut and Executive Vice President Paul Taylor.

Lee Rainie and Susannah Fox are especially thankful for the help and support of their colleagues at the Project: John Horrigan, Deborah Fallows, Amanda Lenhart, Mary Madden, and Cornelia Carter-Sykes. They also greatly appreciate the help on the "Future of the Internet" surveys from Prof. Steve Jones of the University of Illinois–Chicago, a longtime advisor to the Project.

PRINCETON SURVEY RESEARCH ASSOCIATES

PSRA conducted the survey that is covered in this report. It is an independent research company that specializes in social and policy work. The firm designs, conducts, and analyzes surveys worldwide. Its expertise also includes qualitative research and content analysis. The firm can be reached at 911 Commons Way, Princeton, NJ, 08540, by telephone at 609-924-9204, by fax at 609-924-7499, or by e-mail at ResearchNJ@PSRA.com.

ELON UNIVERSITY SCHOOL OF COMMUNICATIONS

Elon University has teamed with the Project to complete a number of research studies, including the building of the Early '90s Predictions Database and the Predictions Surveys on the site, Imagining the Internet (http://www.imaginingtheinternet.org), and a 2001 ethnographic study of families' use of the Internet in a small town, "One Neighborhood, One Week on the Internet" (http://org.elon.edu/pew/oneweek/), both under the direction of Janna Quitney Anderson. For contact regarding the Predictions Database, send e-mail to predictions@elon.edu. The University's Web site is http://www.elon.edu/.

Many people at the University made this work possible. We owe special thanks to Elon's President Leo Lambert and Provost Gerry Francis, School of Communications leaders Paul Parsons and Constance Ledoux Book, University Relations Director Dan Anderson, designer Christopher Eyl, and copy editor Colin M. Donohue.

SURVEY RESPONDENTS

We are thankful for the thoughtful and revealing contributions of the thousands of international participants in the Future of the Internet surveys. Their insights are helping to create significant knowledge about the past, present, and future of information technologies.

Summary of Findings

A broad-ranging survey of technology leaders, scholars, industry officials, and interested members of the public finds that most experts expect attacks on the network infrastructure in the coming decade. Some argue that serious assaults on the Internet infrastructure will become a regular part of life.

In September 2004 the Pew Internet & American Life Project sent an e-mail invitation to a list of respected technology experts and social analysts, asking them to complete a 24-question survey about the future of the Internet. We also asked the initial group of experts to forward the invitation to colleagues and friends who might provide interesting perspectives. Some 1,286 people responded to the online survey between September 20 and November 1, 2004. About half are Internet pioneers and were online before 1993. Roughly one-third of the experts are affiliated with an academic institution, and another third work for a company or consulting firm. The rest are divided between nonprofit organizations, publications, and the government.

This survey finds there is a strong across-the-board consensus that the Internet will become so important to users in the coming decade that the network itself will become an inviting target for attack. By a nearly 3–1 margin, the experts in this survey expressed worry about the vulnerability of the Internet and the likelihood of an attack on the underlying infrastructure within the next 10 years.

Some 66% agreed with the following prediction: At least one devastating attack will occur in the next 10 years on the networked information infrastructure or the country's power grid.

As one expert wrote, "A simple scan of the growing number and growing sophistication of the viral critters already populating our networks is ample evidence of the capacity and motivation to disrupt." Eleven percent disagreed with the prediction and 7% challenged it, including some who argued that they did not expect any attack to be serious enough to involve loss of life or a very long outage.

The Internet will be more deeply integrated in our physical environments, and high-speed connections will proliferate—with mixed results.

There was little disagreement among experts that broadband adoption will grow and that broadband speeds will improve. They also said many more people and objects will be linked online in the next decade. Experts envision benefits ranging from the ease and convenience of accessing information to changed workplace arrangements and relationships. At the same time, a majority of experts agreed that the level of surveillance by governments and businesses will grow.

A full table of predictions and stakeholders' reactions follows in this report. Some of the highlights:

- 59% of these experts agreed with a prediction that more government and business surveillance will occur as computing devices proliferate and become embedded in appliances, cars, phones, and even clothes.
- 57% of them agreed that virtual classes will become more widespread in formal education and that students might at

least occasionally be grouped with others who share their interests and skills, rather than by age.
- 56% of them agreed that as telecommuting and homeschooling expand, the boundary between work and leisure will diminish and family dynamics will change because of that.
- 50% of them believe that anonymous, free, music file sharing on peer-to-peer networks will still be easy to perform a decade from now.

At the same time, there were notable disagreements among experts about whether Internet use would foment a rise in religious and political extremist groups, whether Internet use would usher in more participation in civic organizations, whether the widespread adoption of technology in the health system would ameliorate the most knotty problems in the system such as rising costs and medical errors, and whether Internet use would help people expand their social networks.

Finally, the experts were relatively unconvinced about two suggested impacts of the Internet related to democratic politics and processes:

- Just 32% of these experts agreed that people would use the Internet to support their political biases and filter out information that disagrees with their views. Half the respondents disagreed with or disputed that prediction.
- Only 32% agreed with a prediction that online voting would be secure and widespread by 2014. Half of the respondents disagreed or disputed that idea.

In the emerging era of the blog, stakeholders believe the Internet will bring yet more dramatic change to the news and publishing worlds. They predict the least amount of change to religion.

Asked to rate the amount of change that is likely in a variety of institutions in the next decade, the Internet experts predicted the most radical change in news and publishing organizations and the least amount of change in religious institutions. They also predicted large-scale change to educational institutions, workplaces, and health care institutions. They believe that families and communities will experience change, but not as much as other social arrangements. Here are examples of experts' reflections:

> "Connections across media, entertainment, advertising, and commerce will become stronger with future margins going to a new breed of 'digital media titans'…Well-branded innovators such as Google and Starbucks have a chance to build all-new new distribution models tied to ad revenue and retail sales."
>
> "Health care is approximately 10 years behind other endeavors in being transformed, and will experience its boom in the next 10 years."
>
> "Government will be forced to become increasingly transparent, accessible over the Net, and almost impenetrable if you're not on the Net."
>
> "Digitization and the Internet make for a potent brew…TiVo kills the commercial television format. Napster, Kazaa, and iPod kill the 'album' format. In the future, everyone will be famous for 15 minutes in their own reality show."
>
> "Hyperlinks subvert hierarchy. The Net will wear away institutions that have forgotten how to sound human and how to engage in conversation."
>
> "The 'always-on' Internet, combined with computers talking to computers, will be a more profound transformation of society than what we've seen so far."
>
> "The next decade should see the development of a more thoughtful Internet. We've had the blood rush to the head, we've had the hangover from that blood rush; this next decade is the rethink."
>
> "The dissemination of information will increasingly become the dissemination of drivel. As more and more 'data' is posted on the Internet, there will be increasingly less 'information.'"

Stakeholders are in both awe and frustration about the state of the Internet. They celebrate search technology, peer-to-peer networks, and blogs; they bemoan the slow rate of institutional change.

We asked the experts to tell us what dimensions of online life in the past decade have caught them by surprise. Similarly, we asked about the changes they thought would occur in the last decade, but have not really materialized. Their narrative answers can be summarized this way:

- *Pleasant surprises:* These experts are in awe of the development of the Web and the explosion of information sources on top of the basic Internet backbone. They also said they were amazed at the improvements in online search technology, the spread of peer-to-peer networks, and the rise of blogs.
- *Unpleasant surprises:* The experts are startled that educational institutions have changed so little, despite widespread expectation a decade ago that schools would be quick to embrace change. They are unhappy that gaps exist in Internet access for many groups—those with low income, those with lower levels of educational attainment, and those in rural areas. And they still think there is a long way to go before political institutions will benefit from the Internet.

These survey results and written commentary from experts add to a growing database of predictions and analysis of the impact of the Internet.

At the invitation of the Pew Internet & American Life Project, Elon University communications professor Janna Quitney Anderson formed a research group in the spring semester of 2003 to search for comments predicting the future of the Internet dating from the time span when the World Wide Web and browsers emerged, between 1990 and 1995. The idea was to replicate the work of Ithiel de Sola Pool in his 1983 book, *Forecasting the Telephone: A Retrospective Technology Assessment.* Predictive remarks were found in government documents, technology newsletters, conference proceedings, trade newsletters, and the business and consumer press—comments by nearly 1,000 people were logged in the predictions database, and more than 4,200 predictions were amassed. The fruits of that work are available at http://www.imaginingtheinternet.org, and they were also the basis for a book by Anderson titled *Imagining the Internet* (Rowman & Littlefield, 2005). The idea to launch a series of "Future of the Internet" surveys followed the development of the initial database of early 1990s Internet predictions. The surveys illuminate the status quo and establish people's hopes and fears for the future of networked communications. We expect this information to serve as a resource to assess the evolution of the Internet.

Levels of Change That Internet Use Will Bring in the Next Decade

Respondents were asked to do the following: On a scale of 1–10 with 1 representing no change and 10 representing radical change, please indicate how much change you think the Internet will bring to the following institutions or activities *in the next decade*. The results are represented as the percentage of experts and interested members of the public who predicted change that each point on the scale.

	1	2	3	4	5	6	7	8	9	10	Did Not Respond	Mean
1. News organizations and publishing	*	*	1%	1%	3%	5%	12%	20%	23%	33%	2%	8.46
2. Education	*	*	1	3	5	8	16	21	19	24	2	7.98
3. Workplaces	*	*	1	2	7	9	18	22	18	20	2	7.84
4. Medicine and health care	*	1	3	4	7	10	15	20	17	20	2	7.63
5. Politics and government	*	1	3	5	7	11	19	22	14	16	2	7.39
6. Music, literature, drama, film, and the arts	*	2	4	7	10	10	15	19	14	16	2	7.18
7. International relations	1	3	7	5	13	12	17	17	9	13	2	6.74
8. Military	1	3	7	7	14	11	16	15	10	12	4	6.53
9. Families	*	3	8	9	17	15	16	16	7	7	2	6.24
10. Neighborhoods and communities	1	5	7	9	17	14	19	14	6	7	2	6.16
11. Religion	5	14	16	11	16	11	11	7	2	3	3	4.69

Source: Pew Internet & American Life Project, Experts survey, September 20—November 1, 2004. Results are based on a nonrandom sample of 1,286 Internet users recruited via e-mail. Since the data are based on a nonrandom sample, a margin of error cannot be computed.

How Respondents Assessed Predictions About the Impact of the Internet in the Next Decade

	Agree	Disagree	Challenged the Prediction	Did Not Respond
Network infrastructure: At least one devastating attack will occur in the next 10 years on the networked information infrastructure or the country's power grid.	66%	11%	7%	16%
Embedded networks: As computing devices become embedded in everything from clothes to appliances to cars to phones, these networked devices will allow greater surveillance by governments and businesses. By 2014, there will be increasing numbers of arrests based on this kind of surveillance by democratic governments, as well as by authoritarian regimes.	59%	15%	8%	17%
Formal education: Enabled by information technologies, the pace of learning in the next decade will increasingly be set by student choices. In 10 years, most students will spend at least part of their "school days" in virtual classes, grouped online with others who share their interests, mastery, and skills.	57%	18%	9%	17%
Families: By 2014, as telecommuting and homeschooling expand, the boundaries between work and leisure will diminish significantly. This will sharply alter everyday family dynamics.	56%	17%	9%	18%

Network infrastructure: At least one devastating attack will occur in the next 10 years on the networked information infrastructure or the country's power grid.	66%	11%	7%	16%
Embedded networks: As computing devices become embedded in everything from clothes to appliances to cars to phones, these networked devices will allow greater surveillance by governments and businesses. By 2014, there will be increasing numbers of arrests based on this kind of surveillance by democratic governments, as well as by authoritarian regimes.	59%	15%	8%	17%
Formal education: Enabled by information technologies, the pace of learning in the next decade will increasingly be set by student choices. In 10 years, most students will spend at least part of their "school days" in virtual classes, grouped online with others who share their interests, mastery, and skills.	57%	18%	9%	17%
Families: By 2014, as telecommuting and homeschooling expand, the boundaries between work and leisure will diminish significantly. This will sharply alter everyday family dynamics.	56%	17%	9%	18%
Creativity: Pervasive high-speed information networks will usher in an Age of Creativity in which people use the Internet to collaborate with others and take advantage of digital libraries to make more music, art, and literature. A large body of independently produced creative works will be freely circulated online and will command widespread attention from the public.	54%	18%	9%	20%

(continued)

How Respondents Assessed Predictions About the Impact of the Internet in the Next Decade (*continued*)

	Agree	Disagree	Challenged the Prediction	Did Not Respond
Personal entertainment: By 2014, all media, including audio, video, print, and voice, will stream in and out of the home or office via the Internet. Computers that coordinate and control video games, audio, and video will become the centerpiece of the living room and will link to networked devices around the household, replacing the television's central place in the home.	53%	18%	10%	19%
Internet connections: By 2014, 90% of all Americans will go online from home via high-speed networks that are dramatically faster than today's high-speed networks.	52%	20%	8%	20%
Digital products: In 2014 it will still be the case that the vast majority of Internet users will easily be able to copy and distribute digital products freely through anonymous peer-to-peer networks.	50%	23%	10%	17%
Extreme communities: Groups of zealots in politics, in religion, and in groups advocating violence will solidify, and their numbers will increase by 2014 as tight personal networks flourish online.	48%	22%	11%	19%

Summary of Findings

Civic engagement: Civic involvement will increase substantially in the next 10 years, thanks to ever-growing use of the Internet. That would include membership in groups of all kinds, including professional, social, sports, political and religious organizations—and perhaps even bowling leagues.	42%	29%	13%	17%
Politics: By 2014, most people will use the Internet in a way that filters out information that challenges their viewpoints on political and social issues. This will further polarize political discourse and make it difficult or impossible to develop meaningful consensus on public problems.	32%	37%	13%	18%
Democratic processes: By 2014, network security concerns will be solved and more than half of American votes will be cast online, resulting in increased voter turnout.	32%	35%	15%	18%

Source: Pew Internet & American Life Project Experts Survey, September 20—November 1, 2004. Results are based on a nonrandom sample of 1,286 Internet users recruited via e-mail. Since the data are based on a nonrandom sample, a margin of error cannot be computed.

Up for Grabs

Part 1

Introduction

In mid-2001, Lee Rainie, the director of the Pew Internet & American Life Project, approached officials at Elon University with an idea that the Project and the university might replicate the fascinating work of Ithiel de Sola Pool in his 1983 book, *Forecasting the Telephone: A Retrospective Technology Assessment*. Pool and his students had looked at primary official documents, technology community publications, speeches given by government and business leaders, and marketing literature at the turn of the 20th century to examine the kind of impacts experts thought the telephone would have on Americans' social and economic lives. The idea was to apply Pool's research method to the Internet, particularly focused on the period between 1990 and 1995 when the World Wide Web and Web browsers evolved.

Janna Quitney Anderson, a professor of journalism and communications at Elon, formed an Internet research group in the spring semester of 2003. Elon students, faculty, and staff members searched for comments and predictions from 400 experts made during the period between 1990–1995 in government documents, technology newsletters, conference proceedings, trade newsletters, and the business and consumer press. Comments made by more than 1,000 people were logged in a predictions database, and more than 4,000 predictions were amassed. The fruits of that work are available in the "Early '90s Predictions" section at http://www.imaginingtheinternet. org. The predictions database is a useful resource for policy makers, researchers, and students to assess the evolution of the Internet.

This retrospective research effort inspired the survey that is covered in this report. As more and more commentary was examined from the dawn of the Web, it became apparent that it would be useful to return to many of those experts to see what they currently see on the horizon. In September 2004 the Pew Internet & American Life Project sent an e-mail invitation to a list of technology experts and social analysts, asking them to complete a 24-question survey about the future of the Internet. We also asked the initial group of experts to forward the invitation to colleagues and friends who might provide interesting perspectives. Some 1,286 people responded to the online survey between September 20 and November 1, 2004. About half are Internet pioneers and were online before 1993. Roughly a third of the experts are affiliated with an academic institution and another third work for a company or consulting firm. The rest are divided between nonprofit organizations, publications, and the government.

Some Internet luminaries were involved with the survey, including Vint Cerf, Esther Dyson, Bob Metcalfe, Dan Gillmor, Simson L. Garfinkel, Howard Rheingold, and David Weinberger. Other experts who responded to the survey shared only their institutional affiliation, the list of which includes but is not limited to: Harvard, MIT, Yale, Federal Communications Commission, Social Security Administration, U.S. Department of State, IBM, AOL, Microsoft, Intel, Google, Oracle, and Disney, among many others. But some of the best comments came from those who declined to dazzle us with anything besides their ideas. These respondents opted not to identify themselves in any way.

The Pew Internet & American Life Project and Elon University do not advocate policy outcomes related to the Internet. The predictions included in the survey were written to inspire reactions, not because we think any of them will necessarily come to fruition. We chose topics that have come up in our research, as well as some ideas that have recently been in the news.

Some of the predictions were constructed in a way that contained several statements, and it was often the case that experts would agree with one part of the prediction, but not the other. For instance, one assertion they were asked to analyze was: "By 2014, use of the Internet will increase the size of people's social networks far beyond what has traditionally been the case. This will enhance trust in society." Many of the experts supported the first idea in the prediction (the size of personal social networks will grow), but challenged the notion that this would increase the overall level of trust that people had about one another.

In addition to trying to pack several ideas into one prediction, we tried to balance the statements so that there were roughly equal numbers of predictions with "good," "bad," and "neutral" outcomes. Many of the experts were quick to point out that the deployment of technology always brings *both* positive and negative results. Thus, they often reminded us in their written answers that the "good" outcome embodied in the prediction would not be the entire result of the technology change.

After each portion of the survey—each prediction or each question—we invited experts to write narrative responses to the item they had just assessed. We also gave them the option of challenging the predictions we offered, in case they did not agree with the thrust of the prediction or wanted to criticize the wording we had used. Not surprisingly, the most interesting products of the survey are the many open-ended predictions and analyses written by the experts in response to our material, and we have included many of them in this report. Many others are now entered into the Elon-Pew Internet predictions database available at http://www.elon.edu/predictions.

Since the experts' answers evolved in both tone and content as they went through the questionnaire, the findings in this report are presented in the same order as the original survey. The experts were invited to sign their written responses if they wanted to be quoted in the Elon-Pew database and in this report. The quotations in the report are attributed to those who assented to have their words quoted. When a quote is not attributed to someone, it is because that person chose not to sign his or her written answer.

PART 2

INSTITUTIONS

After sharing demographic data in the survey, the experts were asked the following question, "On a scale of 1–10 with 1 representing no change and 10 representing radical change, please indicate how much change you think the Internet will bring to the following institutions or activities in the next decade."

Religion was seen as least likely to change because of the Internet, while news organizations and publishing were the most popular choice for radical change. All of the choices were ranked somewhere between 4.69 and 8.46 on the 1–10 scale. Here's the breakdown: News organizations and publishing (ranked at 8.46 by respondents); education (7.98); workplaces (7.84); medicine and health care (7.63); politics and government (7.39); music, literature, drama, film, and the arts (7.18); international relations (6.74); the military (6.53); families (6.24); neighborhoods and communities (6.16); and religion (4.69).

Survey participants were next asked to write responses to this statement: "In the next decade, which institutions and human endeavors will change the most because of the Internet? Tell us how you see the future unfolding or point us to your favorite recent statement about the impact of the Internet in the future." They followed with fascinating assessments that are extremely revealing of where the Internet stood at that time and where most informed people expected it would be heading.

NEWS ORGANIZATIONS AND PUBLISHING

Mean Score: 8.46 on a 10-point scale
(1 = no change; 10 = radical change)
It is safe to say that many of the experts here have direct connections to news organizations (if only as consumers) and publishing enterprises. Thus, it is not surprising to find that many were speaking from personal experiences as they assessed the evolution of the media environment. One stakeholder wrote, "The most obvious effects on news media are the rise of Weblogs supplanting the public's attentions to traditional news media, and the slow death of newspapers due to erosion of mindshare by online influences such as news Web sites, chat rooms, message boards, and online gaming."

James Brancheau, a vice president at GartnerG2 who analyzes the media industry, wrote, "Connections across media, entertainment, advertising, and commerce will become stronger with future margins going to a new breed of 'digital media titans.' These companies may not come from the traditional value chain; they will be far more aggressive than existing players. The incumbents are not moving fast enough. Well-branded innovators such as Google and Starbucks have a chance to build all-new new distribution models tied to ad revenue and retail sales."

Mark Glaser, a columnist for *Online Journalism Review*, wrote, "One of the biggest changes the Net will bring in the next decade will be a new way of doing journalism, with media companies being watched by the producers of Weblogs and citizen media trying to co-opt their efforts in some way. The Net is one of the last bastions of independent journalism, so media companies will try to dominate online while smaller players work the niches."

Terry Pittman of America Online's broadband division predicted a series of media and entertainment industry impacts, writing, "Ubiquitous broadband network and P2P connectivity, combined with Internet-enabled technologies and behaviors will continue to reshape the media and entertainment industries and the legal frameworks under which they have operated...Legal and technological controls will not be effective; new laws and technologies designed to maintain centralized control of content will fail to stop ingrained sharing behaviors. After the novelty: The percentage of content purchased versus content acquired through sharing will eventually return to proportions resembling pre-Napster after the novelty and binge effects wear off."

> "The Internet has thrown open the floodgates for participatory news and information, allowing individuals to aggregate information from a broad range of sources, truthsquad what they collect, burrow deeper on topics of concern or interest, and take action on that information."
>
> —Jan Schaffer, director of J-Lab, the Institute for Interactive Journalism

ADDITIONAL RESPONSES

Many other survey respondents shared incisive, overarching remarks after reflecting on the future of news and publishing. Among them:

> "We will no longer be able to create technology products in a clean room void of societal interest. The consumer will be heard one way or another." —**Bradford C. Brown**, National Center for Technology and Law

> "The entire concept of information freedom will be profoundly impacted. I do not refer to price, but to the free flow of ideas; to the expectation that information can be located and accessed as never before." —**Michelle Manafy**, editor, Information Today, Inc., EContent magazine, Intranets newsletter

"The Internet has created information demand that traditional publishing technologies are not capable of meeting...Search technology has changed the way information is presented and sorted. Editors have lost the control they traditionally wielded over the presentation and selection of news...Some are predicting the rise of 'citizen journalism' provided by the man in the street using digital technology. I'm not sure most 'citizens' are that interested in being news providers and reporters, but one thing is very clear: The news is becoming interactive. It is no longer a one-way conversation." —**Janice Castro**, assistant dean, director, graduate journalism programs, Northwestern University

> **"Existing powerful channels will diminish, and new online social networks will evolve and deliver news to people much more organically in the course of their daily lives. We are seeing the 'Blogosphere' starting to impact traditional news channels in this fashion—becoming a catalyst for creating and driving the news, speeding up the news cycle, and delivering critical real-time news and information across millions of touch points."**
>
> —Lyle Kantrovich,
> Internet usability expert
> at Cargill,
> also known for his blog

"The Internet allows small units of thinking to access the larger public audience. Language and ethnic minorities, political groupings like Greens or religious fundamentalists, neighborhood political advocates, distinct territories, and even individuals like the Baghdad bloggers or Matt Drudge, can publish and converse with the larger public. This will continue, despite efforts by the media companies to prevent it. There is too much momentum behind the decentralization movement already." —**Mike Weisman**, Seattle attorney, activist in the advocacy groups Reclaim the Media and Computer Professionals for Social Responsibility

"All institutions, all endeavors that rely on the exchange of information will feel the increasing impact of the Internet. The key is to separate the Internet from the World Wide Web. The Internet is truly the revolutionary delivery vehicle; the Web merely an early indication about how looking for information, finding information, publishing information, sharing information, selling information, and even defining information will change in the future." —**Howard I. Finberg**, director, Interactive Learning, The Poynter Institute for Media Studies

EDUCATION

Mean Score: 7.98 on a 10-point scale
(1 = no change; 10 = radical change)
A number of survey respondents expressed disappointment with the progress made thus far in leveraging the potential of the Internet in formal educational institutions, but they also had great expectations. One stakeholder wrote, "The impact on education will be substantial. For the pre-18s, distant learning and self-paced learning will raise standards and increase diversity. For the post-18s, lifelong learning delivered through CBT [competency-based training] will address skills shortages and have knock-on effects in addressing social issues."

Douglas Levin, a policy analyst for Cable in the Classroom, wrote, "Students are likely to be increasingly dissatisfied with conventional approaches to teaching and learning and to the limited resources available to them in all but the best-equipped schools. In the final analysis, schools would do well to heed the Latin writer Seneca's words, which ring as true today as when they were written nearly 2,000 years ago: 'The fates guide those who go willingly; those who do not, they drag.'"

Moira K. Gunn, host of the well-known syndicated radio program *Tech Nation*, wrote: "Education shall change to an individual-driven endeavor, where knowledge is knowable by impetus of the individual, and the presumed authority of teachers shall erode. A new role for teachers will emerge."

And Webby Awards founder Tiffany Shlain wrote, "The institution of education and the way our intelligence will evolve will continue to change from the Internet. The access to information from anywhere/

anytime will no longer define knowledge based on memory but on 'how' to find information and how to put it into context."

WORKPLACES

Mean Score: 7.84 on a 10-point scale
(1 = no change; 10 = radical change)
One stakeholder wrote, "I believe that the next large impact of the Internet will be in the area of work and organizations. We still work in an industrial era mode of communications. While there will always be a need for physical presence and proximity, advanced ICT [information and communications technology] will create a newer form of decentralized organization that will continue to change the nature of work."

Many other respondents mentioned telecommuting. "The whole notion of 'going to work' is one of the newest of civilization's innovations," wrote Mike O'Brien of The Aerospace Corporation, who is formerly of RAND. "The only reason people 'go to work' in offices is because that's where the paper is. Large companies used to work out of people's homes, viz., the great trading houses of Amsterdam, which were really houses. Lloyd's of London was a coffee shop where the underwriters congregated to make deals; it only became a skyscraper when they needed a place to put the paper. The 'paperless office' is a pipedream, but the mandate for people to congregate physically will be greatly reduced by the Internet, which will have as profound a long-term effect on the development of cities as did the automobile."

> "The Internet will impact how people work in the same company or location, and also change the economics of sharing across companies or borders, not just changing the workplace, but even changing economies."
>
> —Joshua Goodman,
> Microsoft Research

And Mike Botein of the Media Center at New York Law School wrote: "Telecommuting already has begun to happen in a big way in white-collar professions. For better or worse, most research—by academics, as well as students—takes place online. There is at least some possibility for expanded informal 'publishing' online, though it is not clear how seriously this is taken...Opportunities for interaction with foreign colleagues are much greater, since the Internet obviates problems of both cost and time zones."

Peter M. Shane, author of "The Electronic Federalist," in the book *Democracy Online: The Prospects for Political Renewal Through the Internet* (2004), wrote, "The most radical changes will likely involve the workplace, because of the economic incentives involved, and processes of artistic creation, because the Internet is such a fabulous new medium of creation and distribution."

ADDITIONAL RESPONSES

Many other survey respondents shared incisive, overarching remarks after reflecting on the future of work. Among them:

> "Business will continue to both drive change online and be driven by it. Medicine will utilize new technologies in perhaps the most dramatic way—in saving lives, extending lives, and making lives better. And as those worlds evolve, they will naturally bring along politics and government, the media, and educational institutions—those who study and respond to these trends as a matter of course." —**Brian Reich**, Internet strategist for Mindshare Interactive and editor of the political blog campaignwebreview.com

> "As an ever-increasing share of transactions become digital a whole host of functions now performed by paper, phone, and in-person, many of them by middlemen, will become digital, lowering costs and freeing up labor. To take just one example, buying a home could be transformed from an expensive paper and person-intensive process to an inexpensive and streamlined digital process." —**Rob Atkinson**, Progressive Policy Institute (previously project director at the Congressional Office of Technology Assessment)

"The greatest change is likely to be on individuals and the way they perform their work. A growing segment of business needs no longer be at a specific location...The impact on the individual, where he/she lives and works, will alter the structure of our cities, the environment, and much of our society."
—**Ted Christensen**, coordinator, Arizona Regents University, overseeing development of e-learning at Arizona's three public universities

MEDICINE AND HEALTH CARE

Mean Score: 7.63 on a 10-point scale
(1 = no change; 10 = radical change)
Daniel Z. Sands, a primary care internist, assistant professor of medicine at Harvard Medical School, and chief medical officer of ZixCorp, predicted that the Internet will continue to be a transformative influence on health systems. "The Internet is changing health care in many ways," he wrote. "It changes the way clinicians communicate with one another, including consultation specialties. It transforms the way patients and providers access and share information. It lowers barriers of the paper world, making it possible to patients to see their records online and be more involved in their health care. It offers additional channels through which care can be delivered to patients. Because of this, it will force new models of licensing health care professionals, to permit us to deliver care at a distance. It will also force us to change the way clinicians are remunerated, to include non-visit-based care. The Internet will increasingly change patients' expectations of the clinicians, so that physicians will routinely need to offer services

> "I think public health has the potential to change the most with the widespread dissemination of public health information via the Internet (and eventually to the mass media), the earlier detection of the spread of communicable diseases and the ability to treat people remotely—all increasing significantly in the next decade."
>
> —Charles M. Firestone,
> executive director
> of the Aspen Institute

like e-messaging, instant messaging, videoconferencing, and other online services."

Ted Eytan, medical director of the Web site for Group Health Cooperative, wrote, "Health care is approximately 10 years behind other endeavors in being transformed and will experience its boom in the next 10 years. Consumers can now clearly see the differentiation between what they can get online from their health care organization (little to nothing) and from other institutions, and will demand parity.

"This will serve to overcome current financial barriers to provide e-health services. We will move from patient-physician e-mail to complete medical record transparency, and ultimately a true sharing of the health care experience and the accountability for the diagnosis and therapy between physician and patient. A recent statement of mine is '2003 was the year of the secure message, 2004 is the year of the online lab result.'"

POLITICS AND GOVERNMENT

Mean Score: 7.39 on a 10-point scale
(1 = no change; 10 = radical change)
Cheryl Russell, author of *The Official Guide to the American Marketplace* and *Demographics of the U.S.: Trends of Projections*, wrote that the Internet opens access and levels the playing field. "Power was once reserved for those with a lot of money—major corporations, special-interest groups, and political parties. Now anyone with a computer and an Internet connection can make his or her case to the masses. As the grassroots flexes its muscles, the balance of power will shift—not only in the U.S., but internationally as well."

Cynthia Samuels of the Center for American Progress expressed concern about a divide in the way in which people will choose to consume information. "The relationship between politics and media will continue to change and affect how people learn about and choose their leaders," she wrote. "The leveling of access to information will make some people remarkably well-informed and others remarkably misinformed, and unless we push to train young people in critical thinking skills this could become dangerous."

Mack Reed of the Digital Government Research Center at USC posited a worsening chasm between those who are fully connected and those who do not have full information access. "The digital divide will grow ever deeper," he wrote, "as computer-savvy citizens enjoy the fruits of this development, while nonusers (by reasons of choice, ignorance, or poverty) are left to deal with metaspace government representation that will dwindle as more resources are poured into online solutions."

He continued, "News media, politics, and governance promise to change the most thanks to the all-publishing, all-connecting nature of Internet communications. The most obvious effects on news media are the rise of Weblogs supplanting the public's attentions to traditional news media, and the slow death of newspapers...We can expect the nature of sociopolitical interaction to change as well, potentially changing the way prospective voters make up their minds—or even how frequently and consistently they vote on any given race or cause."

Women will gain more power, according to Michael Dahan, the leader of technology projects aimed at fostering peace in the Middle East (based out of Ben Gurion University of the Negev, Israel). "Organizations of civil society in countries undergoing varying degrees of democratization will benefit the most. The Internet will have a significant impact in the Middle East over the next 10 years in terms of empowerment of formerly marginalized sectors, particularly for women. I predict less of a change in Western democracies, where certain processes and realities have been imbedded into political and social systems and thus the change will be less."

Dan Ness of MetaFacts, a market-research firm, wrote, "Political and governmental organizations will change the most, as they have changed the slowest so far. In a decade, they will have returned to a more representative role, contrasting with today's misguided elite-biased misinformation, biased today by listening more to broadband users than the narrowband or offline." And one anonymous expert wrote, "Government will be forced to become increasingly transparent, accessible over the Net, and almost impenetrable if you're not on the Net."

MUSIC, LITERATURE, DRAMA, FILM, AND THE ARTS

Mean Score: 7.18 on a 10-point scale
(1 = no change; 10 = radical change)
While change in this area is expected, a vast majority of the respondents didn't address it in their extended elaborations to this portion of the survey.

Fred Hapgood, a writer and cultural commentator, wrote, "Cultural infrastructure will change the most. Alternative media made possible by new technologies will continue to drive change in both the producing and distributing sectors of radio, TV, the recording industry, and film."

An anonymous stakeholder added, "Digitization and the Internet make for a potent brew. Look for continued disruptive change from the new reality of digital photography, digital music, digital video, digital 'film'-making, digital television, digital news, digital books, etc. TiVo kills the commercial television format. Napster, Kazaa, and iPod kill the 'album' format. In the future, everyone will be famous for 15 minutes in their own reality show."

Another wrote, "The area that will change the most will be arts and entertainment. The ability to receive real-time music and video over the Internet, or downloaded content, will radically transform business models for TV and movies as it is already doing for music. It also will continue to change the relation of the public to artists as it has through fan sites, remixes, and other Internet-based phenomena."

International Relations

Mean Score: 6.74 on a 10-point scale
(1 = no change; 10 = radical change)

One stakeholder who chose to remain anonymous wrote, "The Internet increases options and possibilities. It connects people to each other and increasingly things to each other. The world will get a nervous system, and that is a big deal."

Another wrote, "After spending 3 months in rural South East Asia this year, I think the most interesting developments that will come from the Internet will be from the developing nations. We will see this in every category from virus writing, to online gaming, to offshore tech support and coding. I think the Internet will be become increasingly less English-language-centric."

William Stewart of LivingInternet.com said "globalism" will be the biggest impact leveraged by the Internet. "The global distribution of information and knowledge over the Internet at lower and lower cost will continue to lift the world community for generations to come," he wrote. "...A better-informed humanity will make better macro-level decisions, and an increasingly integrated world will drive international relations towards a global focus. Attachments to countries will marginally decrease, and attachments to the Earth as a shared resource will significantly increase. Communities:... Local communities will organize in virtual space and take increasing advantage of group-communication tools such as mailing lists, newsgroups, and Web sites, and towns and cities will become more organized and empowered at the neighborhood level. At the same time, communities will be as profoundly affected by the capabilities the Internet is bringing to individual communications, providing individuals in the once-isolating city the ability to easily establish relationships with others in their local area by first meeting in cyberspace...Internet applications will change expectations of geographically oriented community organizations, and provide increasingly wide choices."

Moira K. Gunn, host of *Tech Nation*, wrote, "The Internet shall break down the significance of the nation/state as we know it today, and what we will see is the rise of the sovereignty of the individual. We shall also see the rise in impact of groups of individuals in every area, beyond what we have already seen. Changes in entertainment shall respond to the individual, as well as to groups of individual who may or may not know each other."

MILITARY

Mean Score: 6.53 on a 10-point scale
(1 = no change; 10 = radical change)
Interestingly enough, despite the fact that the Internet originally evolved out of research being funded by and conducted for the military, few respondents addressed their extended remarks to its impact on the military. One anonymous survey participant wrote, "The military will obviously continue to develop Internet-related technologies, mostly those of surveillance (except they may call them homeland security or privacy protection)."

FAMILIES

Mean Score: 6.24 on a 10-point scale
(1 = no change; 10 = radical change)
While it is clear that the Internet has had a significant impact on family life, most respondents to this particular survey chose to address its influence on other institutions and activities in their elaborations. Among them are a handful who wrote about the family and Internet use while answering.

Michael Botein, director of the Communications Media Center at New York Law School, observed, "Families, friends, and colleagues hang together much more through the Internet than through the lost art of written correspondence or voice—as seen by the fact that my adult children answer e-mails immediately and phone messages in a week (if at all)."

Bill Warren, vice president of Public Affairs & Community Relations at Walt Disney World Company, wrote, "My window on the Internet is most clear as I watch my children effortlessly roam around the world via the Web. It has already taken them farther than I, as a parent, will ever be able to take them. It's making the world a smaller place, and forever expanding my children's horizons. For that, I am both amazed and grateful."

Joshua Goodman, with Microsoft Research, responded that, in his life, this is the area with the highest impact. "I've been most amazed by the change the Internet has already had in my own family life. My wife and I don't talk any more; we just forward interesting e-mail to each other," he wrote. "The Internet is an incredible medium for sharing and communication, and who do we want to share and communicate with more than our own families? The other people we share and communicate with are our coworkers, and I think the second largest impact will be there."

An anonymous respondent wrote, "The context for family interactions has already changed dramatically. The ease with which children and grandparents can communicate; the ability to message instantly will change the nature of our interfamily relations—and thereby change the dynamics of our personal lives."

NEIGHBORHOODS AND COMMUNITIES

Mean Score: 6.16 on a 10-point scale
(1 = no change; 10 = radical change)
Barry Wellman, head of Netlab at the Centre for Urban and Community Studies and the Department of Sociology at the University of Toronto, wrote, "There will be a move toward networked individualism and away from groups—in work, community, and even the family."

Mike Kelly, president of AOL Media Networks, wrote, "As broadband proliferates the access to information, services, and applications by households and institutions with relationships to households (schools, communities, governments, marketers) will deliver on the promise of the Internet as a personal productivity tool, as well as a communications/information resource."

And Andy Opel of the communications department at Florida State University wrote, "One of the greatest areas of change concerns communities of shared interests. The Internet enables us to find people with similar interests: dogs, diseases, hobbies, musical tastes, etc. These online communities are a vital resource of knowledge that is easily accessible. Connecting people of shared interests and bypassing any gate-keeping filter creates the opportunity for radical communitarian 'open-source' exchange of information, ideas and resources. This has a potential to subvert a business model that isolates people and makes them dependent on fee-based exchanges."

RELIGION

Mean Score: 4.69 on a 10-point scale
(1 = no change; 10 = radical change)
Few respondents chose to comment about changes in the institution of religion. Jordi Barrat i Esteve, of the Electronic Voting Observatory at Universitat Rovira i Virgili, wrote, "The institutions strongly based on information exchange, like international politics, education, arts, or media, will change the most because the Internet is directly linked with the information management. On the other hand, religion is above all a personal field and the Internet is here a tool with less influence."

NONE OF THE ABOVE; ALL OF THE ABOVE

There were quite a few experts whose comments could not be sorted into a specific category. Gary Chapman, director of the 21st Century Project at the Lyndon B. Johnson School of Public Affairs at the University of Texas at Austin, wrote, "Nearly everything will change because of the Internet, and especially as the Internet becomes ubiquitous and all-pervasive, different from the discrete experience it is now on a computer. The 'always-on' Internet, combined with computers talking to computers, will be a more profound transformation of society than what we've seen so far."

Douglas Rushkoff, an author and professor at the New York University Interactive Telecommunications Program, wrote, "The biggest changes, as always with new media, will be metaphorical. It's not that anything in particular that we do on the Internet is so important—it's that behaviors we have online can serve as models for behaviors that change in real life."

Bob Metcalfe, inventor of Ethernet and founder of 3Com, wrote, "Governments will tend toward democracy. Transportation will be refined through massive substitution of communication. The current flight to cities will be reversed. The Internet won't be in schools, it will replace schools. Television channels will be replaced by video blogs and Dan Rather will be dragged off the set."[1]

Peter Denning of the Naval Postgraduate School, Monterey, California, and a columnist for *Communications of the ACM*, wrote, "Greater use of open development processes for technology, as in the World Wide Web Consortium or the Internet Engineering Task Force. Greater separation of people from direct social interaction, leading to decreasing skill in social interaction and more social and organizational problems. Greater offloading of work tasks from organizations to their customers (e.g., do-it-yourself Web sites) with less and less human help or customer service available."

> "Anything that has involved an intermediary will be changed. New kinds of intermediaries will emerge, but the old ones—especially in businesses that have created high margins by being in the middle of transactions—will find their very existences at risk."
>
> —Dan Gillmor, technology columnist, San Jose Mercury News, author of *We, the Media*

Bill Eager, an Internet marketing pioneer, wrote, "We already know that the Internet 'connects' the world. We have been largely wired. Now we are at the point where applications will mushroom for individuals and organizations. In particular, individuals will have 24/7 access to communications, education, and information with the proliferation of a new generation of small, portable, wireless access tools. Full integration of voice recognition will make the Internet both accessible to a larger audience and considerably more human friendly."

Rose Vines, a freelance technology journalist, wrote, "The next decade should see the development of a more thoughtful Internet. We've had the blood rush to the head, we've had the hangover from that blood rush; this next decade is the rethink. Because of that, we'll start to see developments that reflect real human needs—for connection, for privacy, for security, for a sense of belonging. We'll probably also see more attempts at control of the Internet, both by business and governments around the world."

Gary Arlen, a communications and media consultant, wrote, "Global communications and the ability to find and identify ideas. Education will be greatly affected, although this depends on financial/funding and policy decisions. Entertainment will continue to drive access, with new forms of digital interactive games and content arising thanks to the capability of the Internet's connectivity and graphics power. Videophony (video phones) for a variety of business and social uses will be an add-on to the coming VoIP [voice over Internet protocol] onslaught."

And David Weinberger, of Evident Marketing Inc., a Fellow at Harvard's Berkman Institute for Internet & Society and well-known blogger, wrote, "Hyperlinks subvert hierarchy. The Net will wear away institutions that have forgotten how to sound human and how to engage in conversation."

Many other respondents shared incisive, overarching remarks that were not completely tied to any one particular institution or activity. Among them:

> "We will be free to create, share, and organize untethered—using enhancements that in the past were enshrined in revered spaces, devices, and times." —**Christine Geith**, Michigan State University

"A significant percentage of the world's population will have access to the Internet wherever they go within the next 10 years…this presents a radical potential—however, that potential's realization depends on how people and their governments take advantage of that opportunity. With Internet and mobile phone-organized collective actions instrumental in choosing the heads of state in the Philippines, Korea, Spain, and the USA, it is clear that activism and electoral politics are already undergoing radical change. With the emergence of new models of production and distribution, cultural production is undergoing equally radical changes. The education system, the military establishment, and the workplace are full of big institutions that change slowly, but as we have seen in the past 10 years, people find ways around the slowness of big institutions." —**Howard Rheingold**, Internet sociologist, writer, speaker

"The most radical changes will likely involve the workplace, because of the economic incentives involved, and in processes of artistic creation, because the Internet is such a fabulous new medium of creation and distribution. I hope for real-but-more-modest gains in the contribution of the Internet to our democratic life." —**Peter M. Shane**, author of "The Electronic Federalist," in the book *Democracy Online: The Prospects for Political Renewal Through the Internet* (2004)

"The institutions and endeavors most amenable to change are those that are most readily affected by ease of communications and the capacity of ordinary individuals to reach a public audience without intermediation by government or corporate or other ingrained institution. They would include also those that currently require large bureaucracies to process information—such as health care…Opportunities for individual writers and teachers and artists of all kinds to find an audience and thus a livelihood will steadily increase and make the culture much more creative and productive. The trends in peer-to-peer, business-to-business, and business-to-consumer online interactions will continue and move toward seamlessness. The models of eBay and Amazon and Google and Yahoo will continue and become part of the mobile, wireless network with each individual empowered by continuous access to information and a network of workplace and entertainment that is both instantaneous and globe-spanning." —**William B. Pickett**, Rose-Hulman Institute of Technology

"The Internet opens communication channels. Each one of us as a leader must put truly helpful content on those channels and responsibly move it forward in a direction for the benefit of many…the world doesn't need to get more complex. In fact, there can be less clutter and more efficiency in all areas. The world could use more creativity to move us forward through the portal of the next decade." —**Victor Rivero**, editor, writer, consultant, and former editor of *Converge*, an education-technology magazine

"The greatest changes will occur in the arena of trust and human relations. The Internet makes it so easy to obtain, store, and retrieve the most intimate details of our lives that people will inevitably exercise a certain preemptive caution or self-censorship even in their most personal relationships… [When] the Internet and electronic technologies seem completely unremarkable, our notions of privacy and personal space will have been irrevocably transformed…When we reach the point where adults have always understood that their electronic footprints are subject to extensive unwanted tracking and storage, risk-taking behaviors will become rarer, to the impoverishment of our lives as individuals in communities." —**Lois C. Ambash**, Metaforix Inc.

"The Internet will erode individual privacy. It does nothing so well as remember the data that users post. The advent of the Net marks the beginning of wide-scale, self-initiated surveillance." —**Thomas Claburn**, *InformationWeek*, formerly at *New Architect*, *Wired*, and KQED-TV

"There doesn't seem to be any original thought out there anywhere, and I think the Internet bears a responsibility for this that will only increase in the coming years." —**Tom Egelhoff**, smalltownmarketing.com

"The Internet will become even more organic. Wires will fade and the Net will be more like a utility—always on. Every device that computes will be capable of connecting, and generation Z will assume connectivity." —**B. Keith Fulton**, vice president, strategic alliances, Verizon Communications, formerly a senior telecommunications policy analyst with the U.S. Department of Commerce IPv6 Task Force

"Any institution that chooses to ignore or underestimate the "user drives" aspect will suffer adverse consequences...The current regime of copyright protection is an impediment to society moving forward in leveraging technological benefits and furthering creative works. Other impediments are legacy "entitlement" arrangements of traditional media, proprietary exclusionary technologies, regulatory systems that respond more readily to corporate lobbies than demonstrating responsibility to social principles. If there is another lasting lesson that the Internet has brought, it is that change is persistent and unavoidable. It is better to be actively, thoughtfully, and humanly adapting technology than to be creating inertia to resist it." —**Sam Punnett**, president, FAD Research, Toronto, Canada

"Two counter movements [are] pushing stridently against each other...One movement is related to radical democracy, distributed systems, and open source. It is a force for the distribution of power among the many, viral replication of memes and other forms—activity similar to the asymmetrical warfare tactics of Al-Qaeda, to be specific, but also resembling the decentralization of solar energy, of people working with alternative power, going off the grid, defying convention...The opposing move represents something of Marshall McLuhan's media reversals. It attempts to use information technology and networks in a panopticon function, leaning toward increased central control and monitoring, and crunching all activity—both at the granular level and at the statistical mass level. It is an authoritarian backlash to the openness of the Internet and an attempt to put the genie back in the bottle. And it could be successful. The technology supports the success of this movement, but the activity of participants online provides resistance, building distributed forms into the politics of interfaces. Unfortunately, these distributed forms and open interfaces also facilitate panoptic monitoring and may seemingly undo their own advances. It is a bold move, not to fight fire with fire, but rather, to combat control with greater openness instead of going into a secret underground, to avoid becoming the fascist in order to fight the fascists. While idealistically pure, it could be doomed to failure. It is a fascinating struggle." —**Christine Boese**, cyberculture researcher, CNN Headline News

CHALLENGES TO THE QUESTION

We invited experts to challenge the premise of the questions being asked of them. One example of an interesting challenge to this question about the institutional impact of the Internet was this: "Although I have tried to adapt my thinking to this questionnaire and to indicate how much I think each sector will employ the Internet, 'How much change you think the Internet will bring' does not describe the way that the Internet or society works. The Internet does not 'bring' change. The Internet is shaped and developed by these sectors; it is not exogenous to them!"

Philip Virgo, secretary general of the European Information Society Group (EURIM), wrote, "It will take several years for the players to get their acts together and reengineer mass-market access products and services for reliable and safe use by ordinary human beings. It will then take several more years to overcome the growing backlash. Radical change will, therefore, only begin to happen towards the end of the decade. It need not be that way, but the current state of denial is that it looks as though it may. P.S. One of my uncles had a mobile office in the early 1950s (World War II army surplus wireless equipment) and I was using non-Internet e-mail (IP Sharp time-sharing service) in 1977. The pace of change to date has been greatly exaggerated."

Daniel Kaplan, founder and CEO of France's Next-Generation Internet Foundation (FING), disputed the approach to the question. "Actually, I do not believe that institutions and human endeavors will change 'because of' the Internet," he wrote. "I do not believe in technological determinism. To me, the Internet is a tool, a catalyst, and not the cause. Technology never emerges independently from its social context; some technologies emerge, others, whose scientific qualities were at least as good, do not; in fact, we collectively choose the technologies which we (subconsciously) believe will enable us to live the life we have chosen for ourselves. The Internet...is the tool for maintaining or recreating social and functional links in an increasingly individualistic society, where everyone's rhythm is disconnected from everyone else's. It is basically a tool for resynchronization, or for managing our independence without transforming it into loneliness. Therefore, the answer to your question is: Everything and nothing [will be most transformed]...However, it seems clear that all activities which can be entirely digitized, from creation to distribution, will change the most. That goes for news, entertainment, and many services."

Negative Impacts

A few respondents went against the grain and predicted that the world would not necessarily be better off because of the Internet.

One anonymous survey participant wrote, "The dissemination of information will increasingly become the dissemination of drivel. As more and more 'data' is posted on the Internet, there will be increasingly less 'information'...This will affect everything from politics to science/pseudoscience to education. The only vestige of hope may be in the development of integrity, whether mandated by law, developed by private labels, or in the most unlikely of scenarios, the nascence of personal integrity."

Another anonymous respondent wrote, "The Internet won't change most institutions and human endeavors too much, because it's increasingly a cesspool of spam, porn, phishing, and other distracting and annoying commodities, discouraging more intensive and productive use."

A third anonymous Internet critic wrote, "The largest impact will be negative; terrorist groups will figure out how to use (and disable) the Internet, with grim consequences."

Finally, one respondent wrote, "The digital divide will further divide the haves from the have-nots. Even as stationary terminals are being made available in public places, so are the main uses of the Internet moving to high-bandwidth portable devices."

ADDITIONAL ANONYMOUS COMMENTS ABOUT INSTITUTIONAL CHANGE

The following additional contributions to this discussion of the question, *"Which institutions and human endeavors will change the most because of the Internet?"* are from predictors who chose to remain anonymous.

> "Several trends will shape the next 10 years: the extension of the Internet beyond the PC to reach the sensors, actuators, and other embedded computers, the continued incorporation of online information into sectors of society, and the completion of the 'always connected, anywhere' transformation of society...A major debate over the next 10 years will be the struggle over who owns and controls the knowledge of where people are. 'Location-aware' computing can be a lifesaver, or a tool for delivery of new sorts of spam and advertising. The whole concept of the media, what is news, who produces it, and why, will continue to change. This will greatly shape politics, public opinion, and social activism—for both the bad and the good."

> "There's almost no limit to the potential for change. Publications and information-based industries have already been radically transformed, and more traditional industries are seeing their information-based components moved entirely online."

> "Soon being offline will not be an option. As more and more people get on the Internet, more businesses will be there to provide services and to troll for customers. There will be huge demand for security, wireless access, and entertainment. Advertisers will continue to flee print and broadcast media, fracturing that market and forcing them into niches. When everything is available to everyone at the same time there will be no dominant killer-advertising channel."

> "The military, health and medicine, and education will change the most, primarily because each area (a) has strong economic/social/political pressures that will drive change, [and] (b) are relatively cohesive institutions that are capable of executing on

strategic change. I expect wireless networking and decentralization and more participation/control from the grassroots will be at the heart of a lot of change."

> "The Internet will...have a large impact on police agencies, as organized crime and terrorist groups leverage the Internet to victimize millions. By the end of the decade, losses from Internet-related crime and terror will exceed losses from all natural disasters."
>
> —Anonymous respondent

"The Internet has already revolutionized the way educational institutions work ([e.g.,] how we conduct research, how classes are taught, etc.). The workplace has been profoundly impacted by the Net: written memos gave way to e-mail messages, [and] noncollocated team members keep in touch every day through e-mail and instant messaging. News institutions are a bit lost as they start to figure out what to make of bloggers and their newfound power to impact readers."

"Several institutions and human endeavors have already leapt ahead in using the Internet (families have been significantly impacted in terms of generational use of the Internet and what it enables, e.g., IM [instant messaging]; workplace environments are impacted in terms of the Internet, but more likely to be impacted in terms of extranets and intranets). Other institutions are slower to adapt new technologies that are developed as a result of the Internet—take for example the adoption of DOI by the publishing industry, or even the ability to have an integrated patient record in the medical field. What the Internet enables will impact all groups—some groups are slower to adopt technology than others. It is also important to take into account the trends in the intranet, extranet, and other communication-based technologies."

"The most change listed is media/news. The application that will make the most change is RSS [really simple syndication, a family of Web feed formats used to syndicate frequently updated content]. Previously, the news Web site, even though virtual, carried significant value to the individual seeking information. Individuals were not apt to go to multiple sites to get diversity of news. They will continue to not go to multiple sites, but with RSS, diversity of news will be brought to them. This has almost the greatest potential for radical change. All of a sudden, small publishers will have compelling means for distribution. But the RSS readers have to get better."

"[Key issues will be] anything concerning intellectual property and information dissemination, marketing, consumer expectations and interactions with products, brands, and entertainment—who can publish/disseminate content—and what that content will be."

"The assumption that there is 'an' Internet is fascinating. I look at the recent takeover of the Orkut.com site by Brazilians as the most exciting thing happening. It demonstrates how with more people able to participate in ICTs that it will no longer be limited to English and upper-middle-class uses and values."

> "The scope and speed of the global economy, as well as its regulating mechanisms, will create a data tidal wave that will overwhelm existing comprehension mechanisms. Entirely new technologies and societal mechanisms will need to be developed to process data into information (and who knows if wisdom will follow)."
>
> —Anonymous respondent

"The impact of the Internet on today is not understood and we are witnessing the birth of 4th-generation computing. The invisible network revolution. Evolution of the relationship of human and machine…from centralized to decentralized to distributed to morphological structures. Hybrid networks that evolve. Structures that evolve. The Internet has [been] and is changing the flow of people, capital, and information, thereby introducing structural transformations in the institutional fabric of the world. We are talking about an evolution of cyberspace and the relation between the virtual and the physical and the logical. Birth of new worlds, new languages, new processes, techniques, and knowledge. The birth of the Cybernetic Age and the death of notions of industrial and information ages. The Info Age is part and parcel to the Industrial Age. Industrial and Information Ages are about knowledge accretion. Cybernetics is about transdisciplinarity, process automation, and convergence and knowledge creation. We are on the verge of a new renaissance. Science, art, and design! Architects of the future will understand the Internet as the platform for a global youth boom. We no longer can see generations in the same light and driven by segmented histories. The global Youth Movement is networked, cross-disciplinary, cognitively unique, and it is about creating the world we live in...The world we are projecting forward...Our world. Their world. We are immigrants to the future. It's all in our children's hands now."

"Communications is instantaneous and mobile. Society is and will continue to be impacted significantly due to the reality of the technology. The circumstance as catalyst making the impact realized may not have arrived but are present only waiting to be fulfilled. Education is probably the most impacted. No longer does anyone have to attend a class to realize a benefit to an education. Cost factors should be significantly impacted to making education available to everyone worldwide for relatively small costs."

"The most radical impacts will be in areas such as government and public policy as a result of information sharing among the nonelite. Opinions will be shaped by far more—and far less elite—influences than the fairly limited ones in the past, such as major media and government officials. The power of virtual lobbies will continue to grow. However, it is an open question of whether or not the financing for these will remain diffused among the nonelite or be co-opted by corporations or ideological organizations. The other key area of impact will be health care, as the Internet changes the relationship between medical professionals and consumers."

> "We will continue to find new ways of connecting humans to each other and new ways to give over to technology things that humans do now."
>
> —Anonymous respondent

"Organizations and functions that require large numbers of people to actively communicate with one another are more Internet-driven than those driven by passive interactions. Competitive advantage among nations, companies, and peoples will be among those who can apply future technology to their basic needs and infrastructure."

"Business will change the most as companies use the Internet to link themselves with suppliers, distributors, and customers. Governments will be more responsive to their constituents. Education will be increasingly freed from the walls of the classroom. Physical presence will not disappear; in fact, it will be more valuable than ever, because people will not have to 'be there' but will do so only when they choose to."

"In the next decade, these contributing factors: faster Internet; cheaper, faster computers; better mobile devices; better Webcams [and] microphones; [and] cheaper peripherals ([e.g.,] printers, DVD burners, LCD screens, [and] portable storage) will get more people to access the Internet in richer ways in a more affordable way. Also, China, India, Brazil, etc., coming online will change the overall Internet user demographics. These 'enabling' factors will have revolutionary technology advances in communications, payment infrastructure, and information dissemination that will further improve efficiencies in various industries that have a lot of middle men, wiping out established players in medicine [and] entertainment. The Internet will improve the quality of life for a lot of people (affording more items) but will complicate people's lives (artificial necessities) in that overall life-satisfaction ratings could go down."

"The ability to keep in touch with nonlocal family members is affirming, particularly for those with young children and grandparents far away. On the other hand, the focus on the Internet in the home might further contribute to a disconnect from your community/neighborhood/family of place."

"Within 10 years, many more devices will connect (and we'll think back to how quaint it was when we needed a 'browser') in our cars, kitchens, phones, etc. The Internet will continue to be driven by people rather than be constricted by commercial interests. The open-source model will continue to grow in popularity and ease of use…This would improve the efficiency and transparency of everything…The most radical and positive change I can envision is the change in the way people interact with government. If public information, public comment, voter information, planning processes, etc., were to be overhauled so as to make them highly accessible to citizens online, it would go very far toward improving citizen involvement and taking the corrupted old-fashioned media monopolies out of their middle-man roles."

"On information, Internet is increasing the noise-to-signal ratio, leaving fashion and false news a great place. On the other hand, new methods of securing the true from the false will emerge. The source will become more important than the message, as it is in TV. Relationships between individuals will be fragmented more than they are today, implying less commitment in most interactions. This will be balanced in the short term by increased value for family and close-relatives relations. Intimacy will not always mean physical proximity. The commercial side of Web will be comforted in the long run."

"The most changes will probably be in the international/political and business spheres. The Internet has shown itself to be really useful to people seeking to bring information into otherwise tightly controlled societies, or to spread information and propaganda. Terrorists' use of the Web to display their captives comes to mind. In the business sphere, I am seeing a slow but sure decrease in the need for face-to-face meetings or shared workplaces; as of this year, our organization is heavily using Net meetings to save travel expenses or time lost in getting to other campuses. CDC [U.S. Centers for Disease Control] is also supporting more telecommuters, who connect to work via the Internet. This phenomenon is only going to get more common, and this is a HUGE change in the way business is done."

"The Internet has placed the power of information in the hands of the masses. While gatekeepers, such as big media, still exist and will continue to exist, the flow of information is much freer, especially as tools, such as iMovie, have allowed people to create their own media. Entertainment, media, and commerce have been most effected and will continue to be effected as people search out their own truth. From a cybernetics standpoint, we have moved from a hierarchy to a circuit—almost as if the structure of Internet is changing the structure of communication and society itself."

"Person-to-person communication will be the first where various technologies ([e.g.,] IM, voice, data, etc.) converge. However, there will be some negotiation of this, as some people like to be contacted immediately ([e.g.,] cell phone) and some prefer to answer at their leisure ([e.g.,] e-mail, IM). Spam and spim will have to be eradicated or sufficiently curbed for this point to be reached."

"The information-anywhere-anytime future we are fast approaching will heighten the divide between the haves and have-nots. Information is power. Governments also will be transformed by the instant reactionary and amplificatory effect the Internet has. Government adjusted to TV by polishing charisma over substance. The next revolution is already underway, and sacrifices substance completely to rule the infomoment."

> "I see a two-pronged development. The first is the progress of virtual presence from today's fictitious game avatars and two-dimensional business teleconferences to a subtle, nuanced, and authentic representation of people and environments. The second path is the provision of devices and real-time connections so that the granularity of reality is transparently overlain with a Web of context and information."
>
> —Anonymous respondent

ENDNOTE

1. This was written before Dan Rather announced his intended departure date from CBS News.

Part 3

Social Networks

PREDICTION: *By 2014, use of the Internet will increase the size of people's social networks far beyond what has traditionally been the case. This will enhance trust in society, as people have a wider range of sources from which to discover and verify information about job opportunities, personal services, common interests, and products.*

Experts' Reactions	
Agree	39%
Disagree	20%
Challenge	27%
Did not respond	15%

Note. Since results are based on a nonrandom sample, a margin of error cannot be computed.

A majority of those who responded to this question have serious doubts about whether people can really handle larger social networks, whether those networks will be meaningful even if they do expand, and even about the virtues of expanded networks.

Early in the history of the Internet, the first big mailing list on science fiction was created, and users' online conversations turned from shared research to shared interests. Any sense of idyll was quickly broken, however, by the first off-topic and unsolicited messages, later dubbed "spam."[2] Perhaps this pattern of connection and intrusion inspired the mixed reactions to this prediction about the power of online social networks—very few experts agreed with this statement without qualifications. Many respondents agreed with the first sentence of the prediction, but thought the second sentence "did not follow," was "too general a claim," or even was "a horrible generalization" that focused on the tech elite in the United States.

Problems with online security and privacy were mentioned quite often as a deterrent to increased trust in society inspired by the Internet. One anonymous respondent wrote, "I agree that social networks will be larger, but I do not believe that trust will be increased, because the Internet will also bring spam, phishing, worms, and rumor mongering which will mitigate against increased trust."

A few survey participants observed that expanded online networks of trusted commercial contacts have benefited many people, but social networks may not reap the same benefits. One wrote, "Although we have begun to overcome some of the trust issues concerning transactions on the Internet, we are not there yet relative to social interactions. That will take more time. The Internet has not replaced the handshake." Another wrote, "Trusting the seller in eBay creates a particular bond in a particular context, but won't necessarily extend to other areas of trust—particularly between institutions and individuals."

Many were skeptical about the value of an expanded social network. One stakeholder wrote that the Internet "may increase the number of casual acquaintances, but not deep attachments." Benjamin M. Compaine, a communications policy expert and consultant for the MIT Program on Internet and Telecoms Convergence wrote, "People will have a wider range of sources—but most individuals will settle on a small number that they will use repeatedly—much as they use a small subset of the large number of TV networks available already. Impact on trust could go either way (or both)—more sources could equal more differences of info could lead to more confusion and skepticism as easily as more trust."

One anonymous respondent placed this in a larger context by writing, "Like any other mass aggregation of people, peddlers of wares and services, hucksters of all descriptions, and general riffraff will make these larger social networks somewhat less than useful. There will be (and are) benefits, however, for those who can tolerate the virtually milling masses. For example, the Internet is great at aggregating individuals without regard to distance, for example, those who are offering a good or service and those who wish to buy, or patients with rare diseases."

> **"You'll get more information, but much of it will be contradictory."**
>
> —Jonathan Band,
> an attorney specializing in
> e-commerce and intellectual property
> with Morrison and Foerster LLP
> in Washington

Dan Froomkin, a columnist for washingtonpost.com and deputy editor of niemanwatchdog.org, wrote, "The key will be for the Internet to embrace geography considerably more than it has thus far. More Friendsters, and less Freepers. More Craigslist and less eBay. The Internet could be a tool for people to connect with each other in their geographic communities, not withdraw from their geographic communities into a virtual space where the horizons are vastly narrower."

Noshir Contractor, a researcher working under a grant from the National Science Foundation and speech and communications professor at the University of Illinois-Urbana-Champaign, wrote, "It will increase the size of people's 'latent' social networks. They will have access to more of their direct and indirect contacts and greater ability to find out who knows who, who knows what, who knows who knows who, who knows who knows what."

ADDITIONAL RESPONSES

Many other survey respondents shared incisive, overarching remarks to the social networks question. Among them:

"This prediction mixes social and professional networks; the first will not expand dramatically, the second will; it won't be easier to 'verify' information, on the contrary." —**Louis Pouzin**, Internet pioneer, inventor of "Datagram" networking

"I think this has already happened, and people's networks have gotten bigger than they want to handle. We'll be watching how the Internet helps people restrict their connections now, more than increases them." —**Douglas Rushkoff**, author; New York University Interactive Tele-communications Program

"It seems likely that trust in commercial services will grow as people increasingly depend on the Internet. However, though there are spectacular examples of relationships engendered by Internet communications, most Internet relationships connect along limited dimensions. There is still no substitute for dense personal interactions." —**Jorge Reina Schement**, director of the Institute for Information Policy at Penn State University

"I think the first sentence is clearly true, but the second sentence does not follow. The increasing size of social networks may not produce increased social trust, if the Internet also proliferates the circulation of untrustworthy information and practices." —**Peter M. Shane**, author of "The Electronic Federalist," in the book *Democracy Online: The Prospects for Political Renewal Through the Internet* (2004)

> "People will have larger social networks, but most of the relationships will be low-value, low-trust relationships. In general, online acquaintances are generally self-selected for agreement in ideology/politics/religion/etc., so I think that we'll actually see more hardening of radical views, rather than more trust."
>
> —Simson L. Garfinkel, Sandstorm Enterprises, an authority on computer security and columnist for *Technology Review*

"I've seen this dynamic at work on The WELL (www.well.com), a social network that has existed since 1985, and which I joined in 1995. Online culture there is self-governing, defines itself in terms of communities and subcommunities of interest, and is generally tolerant of differ-

ing viewpoints and helpful to those who are new to the environment, or in need of specific information or technical aid."
—**Mack Reed**, Digital Government Research Center, USC

"Thanks to the Internet it's easier to maintain weak ties and spatially dispersed strong ties." —**Barry Wellman**, University of Toronto

"Depends on what one means by 'social networks.' Acquaintances? E-mail correspondents? Advice sources? People one parties with? Maybe our social networks will have more long-distance connections, but they may have fewer direct person-person connections. A person's total number of connections may be a limit of that person's tolerance for connections or time available to spend maintaining them, which would be an inherent limit in how large an individual's social network might become." —**Peter Denning**, Naval Postgraduate School, Monterey, California, columnist, *Communications of the ACM*

> "Trust mechanisms will need to be much more effective and there will be a rising number of reasons to NOT trust what people say and do on the Internet. The increase in the size of people's social networks will happen, but there will also be a profound surge in mistrust."
>
> —Gary Chapman,
> LBJ School of Public Affairs,
> University of Texas at Austin

"Although the evidence is inconclusive—and the social effects of the Internet will change as the electronic media change—I read the existing data as mostly negative. Social networks are weakening, and the Internet is not helping at all." —**Peter Levine**, deputy director, Center for Information and Research on Civic Learning and Engagement, University of Maryland

"I think that there has been a pseudoexpansion of social networks, with people who have just a superficial connection (chat, picture, etc.). I don't feel these are lasting connections on the level of an in-person connection. Witness the use of e-mail in business as simply a bridge to a face-to-face meeting. We haven't been able to replace a connection even among professional colleagues." —**Ted Eytan**, MD, Group Health Cooperative

"We'll see much more frequent contact with people whom we already know over the Internet. But the Internet seems a way to build existing relationships—sometimes just from just brief encounters—rather than start new ones. And a lot more casual meetings will ripen into romantic attachments because of the easy availability of correspondence!" —**Mike Botein**, Media Center, New York Law School

"Generally agree. That said, although we have begun to overcome some of the trust issues concerning transactions on the Internet, we are not there yet relative to social interactions. That will take more time. The Internet has not replaced the handshake. It can produce a broad social network. The question remains: How shallow or deep will that pool of relationships be?" —**Bradford C. Brown**, National Center for Technology and Law

> "People's social networks will be richer and more interesting, but the closest parts of those networks won't be numerically larger—we can only take in about 150 people, virtually or in real space. Trust depends on reciprocity in taking risk, and it's hard to do that without lots of repeat interactions and contextual information. So, richer, better, livelier, but not necessarily bigger."
>
> —Susan Crawford, a policy analyst for the Center for Democracy & Technology and a Fellow with the Yale Law School Information Society Project

"Social networks are not just about size; you also have to consider the quality of your contacts/information. Most people have a limited number of people that they actually trust. I believe it's likely that real-world experience has taught them to be careful. The use of the telephone increased the size of social networks, but I doubt the introduction of the telephone increased trust in society. Having access to a wider range of sources is nice, but there is such a thing as access to 'too much information.'"
—**Robert Lunn**, FocalPoint Analytics, senior researcher, 2004 USC Digital Future Project

"The size and diversity of people's social networks will certainly increase because technology, and the Internet in particular, lowers the threshold for participation in social activities and breaks down traditional social barriers (i.e., race and gender stereotypes are harder to advance online if the identity of the participants is unknown. But the size of someone's social network is not necessarily a good thing. The Internet makes us lazy, it encourages us to stay home in front of our computer instead of going to the local watering hole or community center to engage with people directly." —**Brian Reich**, Internet strategist for Mindshare Interactive and editor of the political blog, campaignwebreview.com

"The benefits will be instrumental—that is people will have better information in the choices they make, but they won't trust society as a whole more." —**Mark Rovner**, CTSG/Kintera

"I agree, except for the part about enhanced trust. The networks will expand, but we will also have to be more selective in evaluating the information received through those networks. The 80–20 rule of information still applies—even if the total amount of information increases." —**Ken Jarboe**, Athena Alliance

"Social networks are like telephone numbers. There are limits to how large they can be, based on time and personality. Some people have large social networks; some people don't. I don't think that will change. However, the Internet may well change the composition of social networks. For example, college-aged students have been able to maintain relationships with high school friends much more easily than was once the case. So, their social networks are not remade in college, as was once the case. How such a change might affect where people choose to live after college, etc., will be interesting to track." —**Stanley Chodorow**, University of California–San Diego, Council on Library and Information Resources

> "Ten percent of the world may be on the Net but so are 10% of the world's criminals and they are using it to automate old crimes and invest in new ones. Unless and until law enforcement and security catch up they will undermine and destroy trust."
>
> —Phillip Virgo,
> secretary general, EURIM

"I agree with this prediction and think it has already happened. In my life, the Internet has connected me to many people around the globe, from many different walks of life, and cultures. I have been able to use these contacts for research and writing opportunities, and have been recruited for a variety of different career opportunities. I believe these contacts and opportunities for collaboration will only increase as the electronic connections afforded by the Internet increase." —**Gary Kreps**, George Mason University, National Cancer Institute

"I generally agree, but would offer these thoughts: The more ethical an individual is, the less he or she needs regulations and such. A tiny population of abusers should not provoke penalties for many, but with one's own vigilance and a better understanding of these people, we will keep them and any associated undesired behavior from such at bay, and focus on the majority of positive results that come of technology-assisted social networks." —**Victor Rivero**, editor, writer, consultant, and former editor of *Converge*, an education-technology magazine

"At the same time, I think certain 'trusted sources' will not have the confirmation of the crowd, instead they will be trusted due to their scarcity or connection only to prime nodes within the network." —**Christine Geith**, Michigan State University

"Trust is a human experience—experience being the key word here—acquired over time. Trust is instinct-based. For example, we feel repulsed by meat that smells bad, because we'll get sick, or we feel apprehensive about the dark, because we may be hurt more easily. Trust for people will continue to act in the same way. We experience and assimilate that information. We will not rewire millions of years of evolution because of Friendster." —**Lorraine Ross**, VP, Sales, USATODAY.com

"Overall, this is a key factor of my Internet vision. However, depending on deployment factors globally and on control of key nodes, there may be considerable pushback based on privacy, security, and other social factors unrelated to tech capability. Governments may crack down on allowing social access to 'undesirables' or personal concerns about safety or security may restrain some individuals from full usage of this capability. Moreover, as business models evolve for social networking, the 'price of admission' may affect acceptance into the 'best' networks." —**Gary Arlen**, Arlen Communications Inc.

"The trick here is that the idea of what constitutes a social network is changing. In the past, if I met someone at a conference, and we didn't meet again all year, I might not consider her to be part of my social network ([i.e.,] a friend, colleague, or acquaintance). But, if using the Internet I read her blog, or see her in an IM session, or briefly message her, I consider her a resource from whom I can draw. I suspect that if the question is asked, "Is your social network larger?" those who are the heaviest users of the Internet will answer in the affirmative, but it isn't clear to me that this is a useful measure by itself."
—**Alex Halavais**, State University of New York at Buffalo

> "I don't believe that this will enhance trust in society. Rather I feel it will transform our notion of society from that of a group of people brought together by physical space into a virtual space containing everyone who shares a common need or business or passion. As to the matter of trust, I think it is likely that given the ease with which one can employ deception in cyberspace, people's general trust in society may diminish a bit."
>
> —Peter W. Van Ness, principal, Van Ness Group

"The Web is also a great way to spread disinformation and propaganda. The checks and balances that exist within nation states don't exist on the Web and market forces may not prevail. Currently, I am receiving e-mail and links to bank Web sites which ask for information or link me to sites that look, for example, like CitiBank, which are not official Citibank sites. Verification can be difficult and the tracing down of who is responsible for such false information is impossible for the average individual to do. It costs money to send a false catalogue of products through snail mail, but costs far less to advertise on the Internet or through e-mail, so I suspect that we will see a proliferation of fraud." —**Fran Hassencahl**, Old Dominion University

"In the fall of 1999 I attended the Harvard Business School Advanced Management Program (AMP). The class of 170 business executives from around the world has since been in continual contact thanks to one of our classmates acting as an e-mail coordinator. He receives messages from individuals concerning life changes and promotions, and e-mails that to the whole class, which spans the globe. Whether Mike is sending a daily joke, or update on a classmate who has just become a grandmother, or telling us who is about to arrive in one city, or another seeking a reunion, he and the Internet have brought together a group of people to create a virtual community, extending the Harvard experience way beyond the classroom." we will see a proliferation of fraud." —**Graham Lovelace**, Lovelacemedia Ltd., U.K.

"While people's social networks will expand in size, those will also take different shapes. A typical person's social network is likely to become more geographically diverse. Some relationships will be purely 'virtual'—with interaction occurring purely through online channels. Face-to-face, local relationships will also decrease and potentially become 'shallower' and less important in the average person's day-to-day living." —**Lyle Kantrovich**, usability design expert, Cargill, blogger

> "Social networks will increase, but not beyond traditional measures. This is because social networks are not driven by electronics. Electronics can help to sustain relations. Accordingly, we will have more interactions with existing networks beyond what has traditionally been the case. New networks will continue to be a function of where we work, live, worship, and play. The Net will enhance these things, not replace them."
>
> —B. Keith Fulton,
> vice president,
> strategic alliances,
> Verizon Communications

"We have not yet found good ways to verify authority in cyberspace. I also think there is a finite attention space for people. We might get new and better sources of information, but they will replace some of the old ones." —**Mike Weisman**, Seattle attorney, activist in the advocacy groups Reclaim the Media and Computer Professionals for Social Responsibility

"The sizes of social networks are almost certainly a function of some innate capacity in people. The Internet can enable them, but not necessarily enlarge them, nor, necessarily, enhance trust in society. A better view is that the Internet reshapes and relocates social networks. By changing the nature of networking and making it more efficient, it is also changing the mix of people who are effective at social networking. It allows finer niche groups, allows the introvert to succeed where he/she may not have before." —**John B. Mahaffie**, cofounder, Leading Futurists LLC

"It will decrease trust and create a more liminal society, a culture of surfaces that are all suspect, and assumed so, like the hypothetical culture at the crossroads where everyone lies all the time and everyone knows it. The cultural effects of the 'lying assumption' don't necessarily bode a loss of integrity, but rather, they unbind integrity from a speaker's ethos, as in the case where greater truths can be uttered with a pseudonym than would be uttered with one's real name. Connection expands, more communication takes place, but identity and representation, and integrity, are unbound from the communication." —**Christine Boese**, cyberculture researcher, CNN Headline News

> "My social networks may indeed expand through the Internet, and I may come to trust people and society, at least in that milieu. But as I go out physically into a social environment, will my trust in society that grew virtually 'work' for me in the physical space? I can see learning to trust and be engaged with people online, but that is not my personal space. Please don't get too close to me in the elevator."
>
> —Barbara Smith, technology officer, Institute of Museum and Library Services

"This seems to describe the situation today, not 10 years in the future. The problem is to have a broad social network that contains people you care about—an online community of like-minded people (or seemingly like minded). Presumably, in 10 years, that will be more common. Online communities might enhance trust in that online group of people but in society generally—I don't see it would have that effect; in fact the opposite seems just as likely. Myriad online interests with their own communities could easily cause trust to dissipate." —**Michael Neubert**, digital projects coordinator, Library of Congress

"We may well know more people superficially, but we will know far fewer people with any degree of depth or abiding interest. Rather than inspire greater trust, our mile-wide, inch-deep approach to relationships and contacts will engender a growing distrust of others. The greater accessibility of information on others also entails a greater access for others to information about us. The loss of privacy and general intrusion of information and technology will lead to an interesting catch-22 for many people: We will hate the developments, but find ourselves unable to live without them. And we will not be happy." —**Daniel Weiss**, Focus on the Family (Christian ministry)

ANONYMOUS COMMENTS

The following excellent contributions to this discussion of increasing social networks and trust in Internet communications are from respondents who chose to remain anonymous.

"Two points: First, a wider number of contacts does not mean a social relationship exists, so I'd like to challenge the use of the phrase 'social network' because it ignores the relations that form that connection. It is unlikely that people will be able to build larger strong-tie networks—as humans we haven't changed—our cognitive limits and perhaps one might say socializing limits have not changed. Sure, some people may sustain many ties they believe are strong, but I'm guessing these people would have maintained them anyway, regardless of the Internet. Second, as people can check up more on others, we can reasonably expect that many people will object to be checked up on. Thus, I fail to see how being able to check up on everyone is related to concept of trust."

"Agree: By 2014, use of the Internet will increase the size of people's social networks far beyond what has traditionally

Social Networks

been the case. Disagree: This will enhance trust in society, as people have a wider range of sources from which to discover and verify information about job opportunities, personal services, common interests, and products."

"In a cable universe with 400 channels, most people use only seven. That's because most humans do not expand as possibilities do. The Internet may allow for easier social intercourse, but ease will not equal more. We do not verify things now with the tools available. There is no likelihood that because it is on the Web we will use it."

"I agree to a degree. But there is also likely to be a backlash to the 'database nation' concept, as many people seek to protect their personal details. Identity theft will have a serious, chilling effect on this trend."

"The Internet does expand social networks—but it decreases, not increases, trust. While the range of information sources is increased by the Internet, the number of reliable sources is not. Similarly, while social networks widen, the number of trustworthy contacts is not increased by the Internet. However, the Internet does enable people who already trust each other to keep in better contact."

"Social networks will expand. The propensity to reconnect with old friends who will be much more accessible is but one example. See www.legacy.com."

"First, I believe that the size of social network is limited by our capacity to interact with much more than a few hundred people. Second, I observe that the Internet makes it easy for people to congregate with other people just like them, rather to seek interaction with a wide range of sources."

"Once you get beyond the reach of what has traditionally been a social network (15 people you truly empathize with and maybe 100 who you care about) the value of these new social connections become negligible. The fact that someone is tangentially connected to me does not make them any more valuable of a contact than someone I don't know."

"I agree with most of it. What I challenge is the 'trust' component. Already, there is evidence of a backlash by some people against the Internet who would contend its content and messages are too easily manipulated."

"People are already on overload. Unless better management and security tools emerge, people may find themselves withdrawing as well."

"By 2014, the elementary children today will be entering the work arena. Those who have computer access in their homes will already be so Internet-savvy that the smart ones will have information on the companies, the people, and the job opportunities in the marketplace. These same children do now and will continue to communicate using Internet tools: cell phone text messages, palm pilots, laptop devices, IM, etc. Much social activity for the Gen X and Millennial (Gen Y) groups is currently being done—meeting on the Net and dating/marrying etc. This will continue involving foreign country connections as well."

> "Liability concerns and information security will severely limit the amount of verifiable information available over public networks. Thus, trust will continue to be diminished, giving way to a cynical view of Internet information and social interaction. Until some paradigm of trust and verifiability is established, social, political, and financial interactions will be limited to role-playing, generalities, and liability-limited transactions."
>
> —Anonymous respondent

"The typical size of a social network of a given person has been estimated very differently, but it's generally been within hundreds, typically. Enter the Internet. I am an introverted engineer. I don't remember people's names very well. I don't go to parties. I leave any reception drained. I should have a smaller-than-average social network. Yet, my address book contains more than 1,200 people, and this is only a fraction of the people I know and can reach out to. Whether this will enhance trust in society is quite questionable."

"The capacity of one's social network will remain the same (the Internet will not change human capacity for intimacy and trust) as will many aspects of its diversity. Geographically, it will change, as we are more able to build relationships with virtual groups and distant individuals."

"It has already happened, so I cannot imagine that using the Internet to do everything from search jobs, find like-minded people, etc., will decrease. In fact, as Wi-Fi and mobile devices become more common and are better made (i.e., easier to use) social networks will only increase. In fact, I suspect that face-to-face time will be radically reduced in favor of digital meetings (with avatars, etc.)."

"There can be no doubt that the use of contact lists and bookmarks is enhanced by the ease of collecting and storing them through PCs on the Net. It's simply easier and faster to gather and verify information today, and it will only become more refined by 2014."

"I agree with the first statement, but not the statement on trust. Unfortunately, we are all still human, and along with the addition of social networks and information flow will come an increase in the amount of untrustworthy information out there, as well. We will have to become skilled at recognizing the trustworthy information from the non-trustworthy information."

"Social networks will change in terms of geography but not that much in terms of size. Personal relationships take time and the Internet doesn't change the amount of time given to relationships. There will be more information available and more dis- and misinformation. Trust will not be enhanced."

"People will certainly have an opportunity to ask others their opinions. They can do that now via chat rooms or message

> "Information overload is a problem; however, work with kids has already indicated that hand/eye coordination and the ability to take in and process more information [are] evolving. That and the fact that Internet search and contextualization continues to improve makes me think this will be true. Individuals and the tools they use are evolving to make people more efficient at taking advantage of access to more people and more information."
>
> —Anonymous respondent

boards. The lack of governance in these rooms and on these boards leaves substantial doubt in the value of the information gleaned."

"As for the development of social networks, it is difficult to fathom the notion that notes hastily sent electronically truly build the bonds required for socialization. Those that believe they do are either remarkably optimistic, or believe that the very nature of socializing will change. If the latter, then there is no quarrel, but then the prediction is merely an exercise in semantic gymnastics."

"Even though we may be able to draw on even more resources than today, we will still have selective perception, suffer from information overload, and stick to whom and what we know, mostly. Today, compared to 10 years ago, there is vastly more information accessible, but most people do not make use of it and, if anything, the widening of our networks and access to everything has decreased trust rather than increased it."

"The cost of sorting through the cheap and rapid communication allowed by the Internet will be too much for the individual to incur. As a consequence of this, numerous errors and misrepresentations will erode social trust."

"Increased information flow tends to widen knowledge gaps; and it seems clear to me that concerns about privacy, identity theft, and fraud will not diminish very much."

"This is already happening, but there are limits to how 'far beyond' people can manage extended networks. People use the Internet to get advice, find information, learn about jobs and travel, find vacations, etc. In the past, that kind of information seeking used to be more limited—usually advice from a few friends or family."

"Social networks will make it easier to make and maintain connections, but this will not enhance 'trust in society.' Just as the phone improved communication, but did not improve trust in society."

"I agree that the Internet will increase the size of social networks, but I do not think that this will enhance trust in society generally. I do not think this follows. I think people will have to become more wary about what the Internet brings them."

"Minor challenge: It will increase the size of *some* people's social networks *somewhat* beyond what has been the case. There will be a minority with super-extended social networks that are amplified by Internet; the majority will use it to maintain social networks they've made face-to-face, or with a few selected friends they've made in online contexts."

"The prediction is based on several assumptions that may not hold true[:] (1) That a larger social network is a better social network, the number of relationships individuals can manage is finite, and nothing about the Internet changes that. (2) That the Internet's main impact on social networks is to increase the number of relationships, when it seems to me the greater impact is on geographic and temporal limits. (3) That more relationships necessarily lead to more information and more information necessarily leads to greater trust. Maybe it just leads to paralysis via information overload."

> "Social networks should not be confused with social life. People will seek advice from credible influencers. However, this will not necessarily enhance trust. The Internet is already being blamed as the cause of scams, crime, and other social maladies that have always existed. This will foster skepticism in some, as others use it more wisely and build trust."
>
> —Anonymous respondent

"I agree with the fact that the Internet allows people to keep in touch with a much greater number of contacts. However, I disagree that this change will automatically enhance 'trust in society.' More contacts does not mean I will trust the society in which I live any more than I do now."

"People will seek out only sources that confirm their current views and positions, sources recommended by those within their current social circles. This will maintain the status quo."

"The Internet is a place where informal communities of like-minded people can 'meet' and share their views. It will allow, in particular, individuals who felt their views to be isolated ones...and held them somewhat in anonymity...to find a voice with others. This can have dangers, just as in groups like the Posse Comitatus and suchlike. But it is also a democratizing force. It is hard to see what the net result will be overall."

"Social networks will expand, but not radically. People only want to know so many people. People will share more information about products, etc., and improve processes like shopping and job changing, but it won't be revolutionary."

"I think the Internet *is* enabling people to meet people they would not have met otherwise, but I do not think the size of people's social networks is increasing. I think they are simply getting more specialized. People can get in touch with others throughout the world, but they can easily avoid being confronted with perspectives and opinions that do not agree with their own. The information on the Internet is so vast and so specialized that, although they have a wider range of resources at their disposal, they can keep their access to a narrow range of these resources that meet their specific needs and comply with their specific viewpoints."

"The increased size of social networks will also lead to more information having to be processed by any one person. That will tend to either degrade communication or increase the consumption burden of individuals."

"There is an upper limit on the time people can spend cultivating their social networks and maintaining ties, and to the extent people maintain more connections they will be weaker...which is not to say that that won't offer some new sorts of value à la Granovetter and weak ties, but I don't expect it to lead to enhanced trust or social capital in and of itself."

"The Internet fosters anonymity and self-separatism, which does not ultimately foster trust."

Endnote

1. According to "Short History of the Internet" by Bruce Sterling, available at http://w3.aces.uiuc.edu/AIM/scale/nethistory.html.

PART 4

NETWORK INFRASTRUCTURE

PREDICTION: *At least one devastating attack will occur in the next 10 years on the networked information infrastructure or the country's power grid.*

Experts' Reactions	
Agree	66%
Disagree	11%
Challenge	7%
Did not respond	16%

Note. Since results are based on a nonrandom sample, a margin of error cannot be computed.

As Americans have become more dependent on the Internet, there has been a growing chorus of concerns about the vulnerability of the network to physical and internal attack. Denial-of-service attacks have plagued some Web sites. A growing number of computer viruses have spread around the network. A central part of the Internet in

Manhattan was near the September 11, 2001, terrorist attacks on the World Trade Center. And Robert Gates, the director of the CIA in the 1990s, recently said that the Internet is a prime target because terrorists may perceive it as a threat to their way of life.[1]

There was strong agreement among experts from every group in our survey that an infrastructure attack was increasingly likely. If there was anything resembling a consensus agreement in our survey, this concern about threats to the Internet was it.

It was obvious that most of the respondents who agreed with this prediction had several kinds of threats in mind. Some, clearly, are worried about terrorists and about physical attacks on central parts of the Internet infrastructure or cyberterrorist exploitation of vulnerabilities in the system that would allow Internet-based attacks on the computer/Internet systems of key utilities such as electric or water systems or key industries such as banking. Others expressed concerns that the electric power-grid on which the Internet and computers depend would be the target. Yet others seemed to be thinking that the network of networks would remain vulnerable to ever-more-clever and ever-more-virulent viruses, worms, Trojan horses, and other packet-born techno-troubles.

At the same time, a number of experts questioned the word "devastating," arguing, for example, that a network attack is not likely to be comparable to a disaster like a hurricane. Fred Hapgood, a professional science and technology writer, responded, "Not if 'devastating' means something like 'no Internet for 24 hours.' It's way too decentralized for that." Simson L. Garfinkel, an authority on computer security and columnist for *Technology Review*, wrote, "I'm not sure what you mean by 'devastating.' We see roughly one devastating attack every 6–12 months. Do you mean an attack with loss of life?" Another wrote, "If you mean very costly, yes. If you mean a failure that cascades to other segments of society with widespread suffering or loss of life, then no."

> "A simple scan of the growing number and growing sophistication of the viral critters already populating our networks is ample evidence of the capacity and motivation to disrupt."
>
> —Anonymous respondent

Network Infrastructure

One self-declared optimist disagreed with the prediction and wrote, "Technology is not sitting still, and our defenses are continuously improving." Another wrote, "Predictions like this reflect a willingness to accept a conspiracy theory of the world. I am too optimistic to agree."

A few experts challenged the prediction. One wrote, "I believe it implies a static infrastructure that isn't constantly being enhanced and expanded." Another wrote bluntly, "Dumb prediction. There is likely to be a devastating attack on anything big."

At least one expert saw a silver lining in the event of an attack, writing, "The question, though, is how we'll weather it. Maybe it will just cause a holiday where we come, blinking, out into the sunlight for an afternoon."

Additional Responses

Many other survey respondents shared incisive, overarching remarks to the question about the networked information infrastructure and the power grid. Among them:

> "It will happen because such an attack would have devastating effects on a country's economy. However, I see no reason for confining such an attack to the United States. All that is necessary is to show that it is possible and to demonstrate the magnitude of the disruption. After that, all of the governments and businesses will have to spend huge amounts of money to harden the target. That will put an enormous strain on the respective economies, and that is the primary goal of many terrorist organizations. The attack on the World Trade Center was an attack on the economic power of the United States. All military power is actually a form of economic power." —**Robert Lunn**, FocalPoint Analytics, senior researcher, 2004 USC Digital Future Project

> "It's already happened, several times, in the form of maliciously disruptive viruses and worms." —**Reid Ashe**, CEO, Media General

> "When World-Pay is off air for days on end does that count as a devastating attack? When a major hub (and the region it serves is off air for hours on end does it matter that the

fire/power failure was accidental or terrorist? When your ISP [Internet service provider] is off air for hours on end during a series of DoS/DDoS [denial of service attacks and distributed denial of service attacks] is that a devastating attack? Those running 2-hour, just-in-time delivery services, let alone life-support services, cannot afford to rely on the Internet. It is too fragile (physical, as well as logical)."
—**Philip Virgo**, secretary general, EURIM

"If we include economic devastation, it's inevitable that we'll see a number of companies and industries up-ended by cracking and by other (more ambiguous) forms of online activity (like file sharing, which arguably has already had devastating economic consequences). If we mean devastating in the sense of directly causing loss of life or injury, it's much harder to predict. To date there are no recorded instances of cyberterrorism (defined by loss of life or harm to human health), which calls into question all the dire predictions about potential online attacks." —**Alexandra Samuel**, Harvard University, Cairns Project (New York Law School)

> "This is the biggest vulnerability to Western lifestyles. But the growth of grid computing may mitigate the risk to networked information infrastructure. The weak link remains power generation."
>
> —Kate Carruthers, Carruthers Consulting

"This cannot be disputed, and both the network and the power grid will fall victim. In the case of the former, software attack is as effective as attacking the hardware infrastructure. In the case of both, there is a huge quantity of hardware infrastructure and outside plant is particularly vulnerable. It is impossible to be completely secure until after attack. The security of the Internet is too much reliant upon self-policing and private sector companies. Government-enforced minimum standards of security are the answer, but I do not see governments having the will or the means to do this. Attacking physical infrastructure has been proven to be both easy and effective and terrorists have targeted physical infrastructure for as long as there has been such infrastructure. The troubles in Northern Ireland, the Basque region, the Red Army Faction, and the Red Brigades have all demonstrated how easy it is for domestic terrorists to attack physical infrastructure, and foreign extremists with no interest in self-preservation will find it even easier. Minimal in-country support is required. Whilst the USA regards itself under attack, its experience of terrorism is nothing in comparison with Europe; getting hold of firearms and explosives is simple, and its security will be easy to breach." —**Steve Coppins**, broadband manager, South East England Development Agency, Siemens

"These attacks occur daily from advertisers, malware, spyware, and other data-mining techniques. If left unabated, the most influential 'attacks' will be from businesses that prey on the uninformed, slowing computers down with pernicious software and turning Internet exploration into a dangerous activity." —**Andy Opel**, PhD, Dept. of Communication, Florida State University

"When I interviewed John Koskinen, President Clinton's Y2K advisor in 1999, he was working overtime to see that the 'rivets didn't fall out of the Golden Gate Bridge of the nation's technology infrastructure,' both here and abroad, in those jurisdictions in which he could assert any control or recommendations that might be adhered to. Thinking about the consequences of a national or international IT [information technology] infrastructure blackout can be mind-boggling. Being prepared and persistently vigilant can help. There are some bad elements out there. They are a tiny percentage of the general population, and over time, they have consistently wreaked havoc on the rest but need not ruin any forward progress we achieve in humane directions. It's been 5 years since Y2K scenarios washed over, and you're not just now climbing out of a mountain shelter, are you? What we need to keep a second eye on is the effect that such scenarios create in terms of spooking entire populations. Sure, there are wolves, but we need not be sheep." —**Victor Rivero**, editor, writer, consultant, and former editor of *Converge*, an education-technology magazine

"There will be many such attacks—the war on terrorism will never be won and will always be fought." —**Bob Metcalfe**, Polaris Venture Partners, inventor of Ethernet and founder of 3Com

"Government has not taken a leadership role in safeguarding the infrastructure so that security measures are fragmented.

Particularly in the current political climate, I see no reason to expect meaningful change in this area. It is not amenable to the self-coordinated efforts of the private sector. It requires not only governmental coordination, but intergovernmental cooperation. Furthermore, the greatest threats to security are not technological, but human. Creating environments and training programs that discourage lapses in security procedures has not been a priority. I believe it is security guru Bruce Schneier who has pointed out that security all comes down to walls and guards and systems that 'fail well.' Most of our systems have not been built with these factors prominently in mind." —**Lois C. Ambash**, Metaforix Inc.

> "The Internet experiences a multitude of attacks on a daily basis. They come from hackers who enjoy disrupting the flow of information. Some of these attacks have indeed been devastating to the targeted individuals or institutions. However, attacking the flow of information is different from attacking the power grid. The power grid, while vulnerable to attack, also contains robustness due to its immense size."
>
> —Jorge Reina Schement,
> director of the
> Institute for Information Policy
> at Penn State University

"Multiple attacks on the networked information infrastructure over the coming decade will cause many people to disconnect and disengage from today's Internet. This will be due to a wide range of concerns, from privacy and security to erosion of their trust in its availability—even to their boredom and declining interest in needing to maintain breaking systems." —**Dan Ness**, MetaFacts

"I don't think this is possible. Although there are many individual organizations that are poorly protected from attacks, most have good defense in place. The power grid is not directly related to the Internet. The power grid is a national security manner. It has always been subject to attack, and has always been heavily fortified and defended by routing around breaks in the network. In the U.S., specifically, the grid is not one grid, but about six grids that are not interconnected. You could take out one, but you would have to attack more than one to take out the country." —**Mike Weisman**, Seattle attorney, activist in the advocacy groups Reclaim the Media and Computer Professionals for Social Responsibility

"It might occur, but not with necessity. In assessing the accuracy of such predictions, one needs to take into account the agenda that they serve. They might want to increase the alertness of those responsible to guard their systems against possible attacks. And they might also want to nourish public fear, in order to justify more restrictive handling of public liberties." —**Albrecht Hofheinz**, University of Oslo

"If the power grid is attacked, that will not be the Internet's fault, but based on how the power grid is established and managed, such as positioning of mission critical systems on an insecure intranet. The Internet itself does not have any inherent weaknesses that would endanger the power grid." —**William Stewart**, LivingInternet.com

> "The Internet is important enough to attack even now and will be even more significant in the future. Terrorists, particularly of the nihilist type now evident in places like Iraq, will see the Internet as a good target for disrupting the Western economy and society."
>
> —Stanley Chodorow,
> University of California–San Diego,
> Council on Library and
> Information Resources

"I'm concerned that most local governments (city/county level) in the U.S., and around the world, are not cognizant of the need to maintain cybersecurity. Given the interconnectedness of government networks, I can visualize ways in which an attack on a city system could cascade to take out utility or public safety nets regionally or even nationwide." —**Tom Foss**, UNC School of Government, Center for Public Technology

"ELF, Al-Qaeda, disgruntled 'patriots,' enterprise crime groups located outside the U.S., 'because we can' hackers, and (hello FirstEnergy; hello all you corporations whose Web sites have exposed Social Security and credit card numbers and other sensitive data) sheer stupidity within the corporate world and its wholly-owned subsidiary, government, are all suspects in the coming attacks. The larger Goliath is, and the more we rely on him, the better a target he is for sending a message." —**Michael Buerger**, Bowling Green State University, Police Futurists International, Futures Working Group

"This is too tempting a target, used for so many commercial transactions, and there are very motivated, crazy people out there who have demonstrated their intention to disrupt and/or demolish our country. I completely believe it is just a matter of time. I just hope everyone has a backup file!" —**Taryn Tarantino**, MarketSource, an Internet marketing company

"Loss of power or information to tens or hundreds of millions for days or weeks will be psychologically more terrifying than the loss of tens or hundreds of thousands of American lives. The magnitude of power or information failure for days or weeks is so great, it is likely to destabilize the e-economy not to mention bring social breakdown to an advanced information and technology society such as ours. Consider what happened in New York City when the lights went off for a day." —**Stan Faryna**, president, Faryna & Associates Inc.

"I challenge the way you ask the question. The net-worked information infrastructure is not a national infrastructure—for the U.S. or any nation. The power grid of the U.S. may be able to work in isolation from the rest of the world, but the U.S. is unable to meet the demands for all kinds of energy with domestic sources. Thus, power is a border-crossing phenomenon as well. Strange that it seems to be so hard for U.S. citizens to wrap their minds around. So the real question: Will one devastating attack occur in the next 10 years on the networked information infrastructure or an important source of energy? The answer is that both are happening all the time." —**Charlie Breindahl**, University of Copenhagen

> "Forces in commerce and society fear the distributed nature of the Internet and are working diligently to layer centralized control on top of the Internet. It is the centralized structures that are vulnerable to attack and will be the ones to topple. Of course, the rest of the Internet will merrily chug along."
>
> —Scott Moore, Charles and Helen Schwab Foundation

Network Infrastructure

> "With greater centralization of Internet networks and the continued hegemony of software and other technology of a few super-companies (including Google), a single big virus or other security blow will be enough to bring down much of the Internet."
>
> —Bornali Halder, World Development Movement

"Terrorism uses a lot of networks, and as communications will be more and more important, they will attack in this new environment." —**Jerome Jolion**, State of Geneva–CTI

"I think there will be several different types ranging from more sophisticated computer viruses, morally questionable content bombarding youth, electronic 'bank robberies,' as well as attacks on the networked information infrastructure or a country's power grid. Not just one, but many." —**Linda Hurt**, systems analyst, Office of Personnel Management

Anonymous Comments

The following excellent contributions to the question about the networked information infrastructure and the power grid are from predictors who chose to remain anonymous.

> "We have already seen the release of a 'zero day' virus (a virus for which no patch is available) whose aim was the theft of personal financial information. Within the next decade, a 'zero day' virus will be launched, which will compromise the financial data of millions of users within a very short period of time (a few hours or less). Banks will scramble to contain the damage."

> "Well, in a sense I agree. I have seen large attacks already happen, e.g., against the Microsoft servers. But technology is not sitting still, and our defenses are continuously improving. I am an optimist, and I believe that defenses will improve quickly enough to ensure the next attack is not 'devastating.'"

> "I might disagree with the word 'devastating.' I think there will be an attack, but I am not convinced that the impact will be as great as we might fear it will be. There may indeed be

attacks, but I doubt that 'devastating' will be the result. The Net is resilient—it was designed to be so. Now, the power grid may be a different story…"

"Countries such as North Korea are already training hackers for use in the national military."

"Not if we can help it—and we're trying." —*From an engineer at a major tech company*

"As the value of the infrastructure increases, the power to use it or disable it becomes a more politically palpable tool for good or ill."

"That's the next logical step for terrorism. Bring this country to its knees with its cyberinfrastructure vulnerability."

"There is considerable, and growing, redundancy and resilience. An attack will very likely occur, but its severity will be more like a typical hurricane or earthquake—troublesome, repairable, but not 'devastating.'"

"Fundamental Islam understands how the West works and will seek to attack key economic and profile targets."

> "Security, and correctness of implementation more generally, is not taken seriously by the computer industry. Small wonder—taking correctness seriously would increase the cost of everything computer related perhaps by an order of magnitude. Only catastrophic attacks could change the attitude."
>
> —Anonymous respondent

"Hundreds of thousands of attacks happen each week. 'Devastating' means one is successful—one is not likely to lose money on this bet!"

"Terrorists these days are smart and will go for the basic infrastructures. Also, while computers are great and help things work better, we need to not forget how to survive without in the event of such an emergency. We need a traditional non-computerized backup."

"It's an obvious target for terrorists or hackers…and as we know, it will take such an attack for the 'establishment' (especially lazy tech/software companies) to fix the numerous security problems that plague the Net…but ironically, without the Net, it is the small disenfranchised groups, like small cults or terrorists, who will be most harmed…since it provides them with vast power that would otherwise be reserved for massive corporations or governments…if they take it down, they are likely to go down with it."

"Not if the network is designed well and has numerous redundancies. The current mesh network is very robust and was designed to deal with many types of grievous attacks."

"Given the current terrorist context we live in and the interest in hackers to show off their skills, this is inevitable—as is the unfortunate human quality to only fix the problem once it has occurred."

"Depends on what you mean by devastating. If you mean very costly, yes. If you mean a failure that cascades to other segments of society, with widespread suffering or loss of life, then no."

"There will be power-grid failures without attacks."

"We have worked to harden the network. That said, the Internet was built to withstand a decentralized attack. It remains an open issue as to what might happen if the attack was more focused. The country is so reliant on the network from a business and financial perspective it is highly likely that such an attack will be attempted. How effective it will be is another matter. The redundancy that has been created will go a long way toward diluting the potential impact."

"I believe it will take that in order for governments, companies, and organizations to make the needed investment in security."

> "There is no reason to assume that there will be a devastating attack. The Internet has already survived the worst attack (September 11, 2001) on the U.S. Efforts continue to harden the infrastructure. The power grid will also benefit from IP infrastructure, as broadband over powerlines is providing an excellent way to monitor and manage network elements in the power grid."
>
> —Anonymous respondent

"This has been predicted for many years without fulfillment, especially by Richard Clarke. However, the same was true of megaterrorist attacks prior to 9/11. I hope Clarke is wrong this time, but he's probably not."

"Well, it depends what you mean by 'attack' and by 'devastating.' We are just as vulnerable to system failures as in the blackout of East Central U.S., in 2003. I believe similar outages are likely and will create the same kind of chaos. Will they be deliberate? Quite possibly, and as likely to be instigated by mischievous 'hackers' as any politically motivated group."

"By attack, you may mean a technical meltdown (similar to what we've seen in past decades), as well as a hostile assault. I strongly agree that some problems will arise…chaos theory almost assures such an attack in the Internet structure…tectonic plates will shift!"

> "Control networks are not secure and they reflect monolithic, siloed, closed, proprietary, and centralized/decentralized architectures. Control networks are our largest vulnerability in the U.S., in the *world*: critical infrastructure air, sea, land, and space. Fourth-generation computing and regulation can help, but we have to get moving. Industry needs to lead or be compelled."
>
> —Anonymous respondent

"Security is nonexistent."

"I think for those who are interested in carrying out devastating attacks, there are easier, lower tech, and higher media-value avenues."

"I'd modify that to be 'attack or event'—I don't discount devastation as the result of ineptitude or poor planning."

"These kinds of attacks, or at least the nascent form of them, are probably already happening, and we don't hear anything about it for security reasons."

"[It is] inevitable that terrorists will attack the infrastructure, since it is becoming a symbol of Western cultural values."

"The Internet, while it makes our lives so much easier and productive, makes us extremely vulnerable. When your way of life is fully electronic, all one has to do is cut the power and watch us flounder. A recent Yahoo! Internet deprivation study showed how people 'lost' their ability to manage certain life tasks (like going to the phone book to look up a number—give me a break!) when their access to the Net was gone. Extreme reliance on the Net makes us less resourceful as human beings. Beware, this is scary."

"Setting aside the obvious threat of terrorism, in terms of attempts at political or economic disruption, I think it is highly likely that a 'hacker' will launch a significant attack to demonstrate our vulnerability. My understanding is that of some members of the hacker subculture are activists of a kind. They seek to reveal the limitations of the technology in order to improve it."

"Our society's security is built on a high level of trust. This is especially true of our information infrastructure. If bitter enemies (like Al-Qaeda) do not succeed in attacking it, vandals (like hackers) almost certainly will."

"Power systems have already shown themselves to be vulnerable, as seen in the 2002 power outage across the Northeast United States. Communications systems are less obvious fragile, but the increase in users of the Internet, without infrastructure development, suggests to me that this is likely to happen soon."

> "Whenever an unchecked power dominates, marginalized communities take it upon themselves to challenge that authority. The network infrastructure has become such a powerful tool."
>
> —Anonymous respondent

"This will, of course, foster more thought of fault tolerance and redundancy."

"The systems are sufficiently adept to avoid a situation that would be termed 'devastating.' There may be incidents, but the likelier near-term stress will be on the capacity of the systems themselves. This gradual pressure will be a greater force and factor to reckon with than any single, momentary attack."

"My only disagreement might be that I think the 'or' is optimistic. I would predict significant and successful attacks on both. The targets are as symbolically significant as the World Trade Center or the Pentagon. Attempts to thoroughly disrupt, corrupt, or dominate the information infrastructure seem inevitable. The vulnerabilities are too well documented and already seem to invite sociopaths, egomaniacs, and nihilists; how far behind can the terrorists be? Power grids are tied to the nets, why not a combined attack? The power grid is symbolic in another way as well, reflecting the gluttony for energy and other natural resources that is represented by the West. This attack could come from Western ecoterrorists as easily as from foreign sources."

"Considering information infrastructural 'protection' is largely left to the free market private sector, it seems likely that there will be at least some big companies that fail to invest in their network stability...and, certainly the deregulated power grids have shown no great capacity to keep up their technologies for the public interest... they'll let them rot as long as there's more (short-term) profit in leaving them poorly serviced."

> "Attacks on these structures may happen, but 'devastation' is unlikely. I'm still much more frightened of a typical truck bomb, which is much easier to construct and deliver than a complicated attack on a power grid or information infrastructure—something that would cause frustration and annoyance, rather than mass casualties."
>
> —Anonymous respondent

"We already know the terrorists have been thinking about and planning this. Our power grids are based on the 1980s infrastructure, without advanced security updates for the modern Information Age. We are vulnerable. And yet, Congress and the president still cannot agree on an energy policy to upgrade everything."

"Both are too decentralized to make it worth it, and neither makes great press. Blood and gore and death have greater emotional impact. Besides, why attack the U.S.'s power grid when it is so poor as to black out an entire section of the country on its own? (If terrorists could figure out a weak point, sure, but they want carnage.)"

Network Infrastructure

"We have already seen a major problem with the country's power grid; this one wasn't even an attack. Imagine what someone really trying to cause harm could do. As for the network infrastructure, anything that we become increasingly dependent on will be subject to danger."

"It took 20–30 years for the theory of commerce warfare to be implemented from theorization in France to implementation by Germany in World War I. Ten years is about half that and is likely appropriate."

"My understanding is that disruption of the top-level DNS servers could be devastating. If our top-level infrastructure is decentralized and fail-over servers are available, one hopes that disruption would be minimal. Organized crime/terrorists must certainly be considering cyber crime since our society is so reliant on online systems."

"There will be devastating attempts, but they will be prevented by the InfoSec teams."

"The toolsets are available to those who would like to do so. The monoculture of Microsoft environments does nothing but encourage such an attack."

"Commercial vendors have no interest in common security solutions. The monopoly position of Microsoft will emphasize the ease by which IT systems will be attacks or compromised on a large scale."

"Devastating for some, but not all or not for a long time period, however. I believe a power-supply grid attack would likely devastate a region, but if that region is later isolated, the rest would function normally. A network information infrastructure attack that was physical would have similar impact as on the power grid. A 'virtual' network attack could devastate a larger portion of the network, but probably for a shorter time period."

"It is inevitable: The Internet is the Mount Everest of hackers, and terrorists are bound to find it more irresistible the more dependent we become on it."

"Technology is simply moving too fast to allow for the proactive protection of the infrastructure. Only catastrophes bring the necessary attention to 'the grid.' I wish it were otherwise, but it's simple human nature."

ENDNOTE

1. "Ex-CIA Chief Gates Warns on Cyberterror," published by the Associated Press, December 4, 2004, available at http://www.sanangelostandardtimes.com/sast/home/article/0,1897,SAST_4943_3375717,00.html.

PART 5

DIGITAL PRODUCTS

PREDICTION: *In 2014 it will still be the case that the vast majority of Internet users will easily be able to copy and distribute digital products freely through anonymous peer-to-peer networks.*

Experts' Reactions	
Agree	50%
Disagree	23%
Challenge	10%
Did not respond	17%

Note. Since results are based on a nonrandom sample, a margin of error cannot be computed.

The Pew Internet & American Life Project reported in January 2004 that the percentage of online Americans downloading music files on the Internet plunged after the Recording Industry Association of America began filing suits in September 2003 against those suspected of copyright infringement. The number of music downloaders has rebounded to some degree in the past year. In previous research, file swappers seemed indifferent to the copyright status of the music they were sharing and downloading. But the country has gotten a crash course on copyright in the past year and a lot of people who trade files online may believe now that indifference to copyright law is a much more risky attitude.

Internet experts were divided into four different camps when it came to the future of digital property rights: philosophers, pragmatists, hacker devotees, and skeptics. Again, while it might appear from the survey data that there is solid agreement among experts on this assertion, it is more appropriate to say that the expert community is very sharply divided and uncertain about the future of shared digital products.

The philosophers felt that society will change to accommodate the new realities of file sharing. Stanley Chodorow, a historian and university administrator, wrote that, "Over time—longer perhaps than the 10 years asked about in the survey—people will begin to see the Internet as part of society and not as a wild territory beyond civilization where anything goes. As that change of view occurs, people will begin to obey rules that will reduce substantially the theft that goes on now."

Ted Eytan, medical director of the Web site for Group Health Cooperative, agreed, citing historical precedent: "Like the sheet music industry in the early 1900s, a new generation of users will grow up in an area of digital rights management and a new norm of not sharing copyrighted material." A third expert envisioned a different future: "Millennials [children currently in elementary and secondary school] will be so computer/network savvy and so interdependent with one another that they will, without hesitation or conscience, share everything."

Some experts focused on the commercial aspects of the prediction. Bob Metcalfe, inventor of Ethernet and founder of 3Com, wrote, "We must, and therefore will, fix this problem. Private property is too valuable an economic tool." Another expert agreed, but rather more bitterly wrote, "We're headed towards digital prohibition. The bastards are going to win."

Faith in technology and the ingenuity of computer programmers inspired a few experts. One wrote, "I agree because to think otherwise would mean that a draconian clampdown on the legitimate uses of the Internet would have taken place. The Internet is designed to allow computers to communicate directly with one another."

Challengers questioned the assumptions behind the prediction. Robert Lunn of FocalPoint Analytics, a senior researcher on the 2004 USC Digital Future Project, confronted its parochialism: "I doubt China's Internet users are going to have that type of 'digital freedom,' and there are a lot of people in China." Michael Wollowski, a computer scientist at the Rose-Hulman Institute of Technology, wrote, "I want to suggest that 'freely' would not imply 'illegally' in 2014."

> "Many of the folks running the Internet and its infrastructure are deeply committed to preserving this capability."
>
> —Anonymous respondent

A few experts challenged the notion that a "vast majority" of Internet users have the skills to participate in peer-to-peer networks. Eszter Hargittai, a Northwestern University professor researching the social and policy implications of information technologies, wrote, "This is not even true in 2004. Use of anonymous peer-to-peer networks is not as easy and as second-nature to the average user as one may think. Once we study online skills in more detail we will realize that the average user knows less than academics/journalists tend to assume."

ADDITIONAL CREDITED RESPONSES

Many other survey respondents shared incisive, overarching remarks to the question regarding copying and distribution of digital products. Among them:

> "Anonymity is not something we'll continue to treasure. In order to create trust, we need to know something about one another—we need to have persistent (even if pseudonymous) identities. These identities don't have to come from a central

source—indeed, there should be competition for the provision of authenticating information, and lots of nuance with respect to how much identifying information is needed for a particular communication. This sort of thing must not be government mandated. Private ordering will map much more tightly to the groups it serves. So, anonymity may cease to be possible, but pseudoanonymity will thrive." —**Susan Crawford**, a policy analyst for the Center for Democracy & Technology and a Fellow with the Yale Law School Information Society Project

"The music, publishing, and other related industries will realize the benefit of opening their stores up to everyone. The industry will evolve to the point where users want to share publicly, not anonymously, and culture grows collaboratively and openly." —**Brian Reich**, Internet strategist for Mindshare Interactive and editor of the political blog campaignwebreview.com

> "Peer-to-peer is impossible to eradicate technologically or legally, and the industries affected are beginning to learn that they have to figure out how to live with this and make a buck—which outfits like Apple, Sony, and Microsoft seem to be proving can be done!"
>
> —Mack Reed,
> Digital Government
> Research Center, USC

"There will always be independent software developers and hackers who are capable of circumventing any technological obstacles to file sharing. The only question is whether the content-creation industries will be able to develop product offerings and business models that are more tempting than file sharing." —**Alexandra Samuel**, Harvard University, Cairns Project (New York Law School)

"There is a strong movement toward open-source access to information and publications on the Internet. I regularly share my publications electronically and download others' publications. We still have to iron out some of the copyright laws, but I am confident the trend toward greater access to digital products, both scientific and artistic, will continue to grow." —**Gary Kreps**, George Mason University, National Cancer Institute

Digital Products

"Digital rights management vs. piracy (or, alternative phrasing: digital rights management vs. public perception of 'fair use'): It's a technology spiral, much like spammers vs. spam blockers. What one side protects the other will crack; so the battle will continue, at least for a minority of users. But, for the majority of people, the development and expansion of services such as iTunes and new innovations in multimedia distribution will provide satisfactory solutions." —**Rose Vines**, freelance technology journalist, *Australian PC User, Sydney Morning Herald*

"It is equally possible that intellectual property will be over protected and fair-use rights will have eroded." —**Peter Levine**, deputy director, Center for Information and Research on Civic Learning and Engagement, University of Maryland

"Old digital projects, yes. The new stuff will be protected better by virtue of its tremendous size. HiDef TV, for example, cannot be copied with current set-top boxes." —**Douglas Rushkoff**, author; New York University Interactive Telecommunications Program

"Most material will be easily copied, but some material will be locked up. Most likely, it will be possible, but it will also be possible to get caught, and the penalties will be drastic. Hopefully, though, most people will find it easier to simply pay for copying fees, rather than trying to get something for nothing." —**Simson L. Garfinkel**, Sandstorm Enterprises, an authority on computer security and columnist for *Technology Review*

> "The anonymous peer-to-peer networks will only survive so long as the industry refuses to offer consumers a viable alternative. The most successful industry in every technology cycle is the adult film industry. Digital media industry business strategists need to learn lessons from these thinkers, who continue to drive technology innovation as a way to maximize profits."
>
> —Elle Tracy, The Results Group

"I think technical impediments, such as digital-rights-management systems, will prevent this." —**Jonathan Band**, an attorney specializing in e-commerce and intellectual property with Morrison and Foerster LLP in Washington

"The prediction does not say if they will do so legally. I expect that business models will be worked out so that there is plentiful copying and it's all legal." —**Peter Denning**, Naval Postgraduate School, Monterey, California, columnist, *Communications of the ACM*

"In 2014, although users may not specifically use anonymous peer-to-peer networks in the sense of Napster, the widespread free trade of digital products will re-emerge after an aggressive half-decade of failed attempts at copy protection by commercial interests." —**Dan Ness**, MetaFacts

"It will depend on what type of product. Intellectual property will still have an owner. Owners of content will have the ability to use sophisticated digital rights management tools (DRM) to determine who has the access to the content and who does not. However, where rights are granted, people will be able to distribute the content freely and easily through P2P and other networks."

—**Marty Shindler**, The Shindler Perspective, Inc.

"Of course, hackers, hacktivists, and rogue programmers will always find a way to get round DRM-style measures of RIAA and their equivalents and share these methods freely. It will become easier to be anonymous on P2P networks." —**Bornali Halder**, World Development Movement

"Copying and distributing will still occur. Licensing and copyrights will evolve to better address this reality. Corporations will come to realize the value of getting products in people's hands before they make a purchase. Demos and trialware will proliferate, as will 'pay as you go' models." —**Lyle Kantrovich**, usability design expert, Cargill, blogger

"All content once digitalized should be freely distributed to everybody. Information and culture to everyone, please." —**Antonio Coelho**, Globo.com

"This is one of the value-added features of the Internet. It cannot be defeated through software or DRM, and I have never understood why they try." —**Mike Weisman**, Seattle attorney, activist in the advocacy groups Reclaim the Media and Computer Professionals for Social Responsibility

"In another 10 years, I would hope that consumers and the creators of digital products will arrive at a new social contract that eliminates anonymous P2P networks, makes digital products easily accessible and affordable for consumers, and rewards creators for their work." —**Andris Straumanis**, University of Wisconsin–Eau Claire

"Antipiracy measures have been largely ineffective. Organizations like the MPAA and RIAA need to embrace new technology and figure out how to benefit from it rather than denouncing it as evil. There were predictions that the VCR would destroy Hollywood years ago but it did the exact opposite. You cannot stop technology. If you want to succeed, you will have to embrace it and make it work for you." —**Jay Buys**, Fleishman Hillard

"And the entertainment industry will at last have moved to the 21st century, from a product economy to a service economy, thereby making more money than ever while contributing a little bit more (but not much more) to cultural inventiveness and diversity." —**Daniel Kaplan**, France's Next-Generation Internet Foundation (FING)

"However, if barriers are erected high enough or the cost of alternatives is low enough, the scale of use could drop to just those willing to do the work to get around them. In the end, information wants to be free, and no one can prevent a copy being made of a book, painting, song, or movie." —**William Stewart**, LivingInternet.com

"If, by 'freely,' you mean at no charge, I disagree. Too many people have an investment in intellectual property, for any of the major governments of the world to end that. As long as there are copyright, patent, and related laws, many of the owners of intellectual property will want to be compensated for its use, even though information wants to be free." —**Elliot Chabot**, senior systems analyst, House Information Resources, U.S. House of Representatives

Anonymous Comments

The following excellent contributions to this discussion about the copying and distribution of digital products are from predictors who chose to remain anonymous.

"The current copyright battles remind me of prohibition and its eventual overturn. When you have 5 billion people wanting to imbibe information in a certain way...it's the providers that have to change their methods and we are already seeing this occur in news, music, software, etc. They will put up an initial fight, but by 2014, the battle will be won and an accommodation made."

"Copyright and intellectual property issues will tie this up for decades in the developed world, where most of the valuable content originates. Until there are mechanisms for revenue sharing that support peer-to-peer transfer, this won't happen. I don't think it will until we restructure our economic and legal system, which certainly won't happen by 2014."

"We have already seen some test marketing of digital books, and it has not gone very well. It seems consumers actually like to sit with a good book and turn the pages. This will be one of those marketing issues that consumers decide. Music is another story. We have been very wary about changing the copyright system. The major blockade to the distribution of digital products is making that system work in some easy, legal way where gaining the rights does not cost so much that it prices the digital product out of the market. We are not there yet."

> "This comes down to decentralization or centralization. If security measures increasingly lead to centralization and lack of privacy protections on the Web are continuously backed up by increased legal reach through copyright and patent law, I could easily see this as the prohibition of our century. Decentralized licensing leads to a model where information can circulate, creators can be compensated, and existing bottlenecks of control disappear."
>
> —Anonymous respondent

Digital Products

"I am pretty certain that technologies that enable what Grant McCracken calls a 'culture of commotion' will outstretch legal attempts to regulate such circulation, but it is going to be an ongoing battle."

"Peer-to-peer networking is here to stay. What will have to change is the transactional business. Peer-to-peer is a disruptive technology that can only be addressed through innovative means."

"I hope that this prediction turns out to be true, but I doubt it will be. There will be increasing pressure to 'lock down' digital products and to eliminate anonymity. That's what I expect."

"With micropayments starting to make more sense, everything will be plug and play."

"Encryption technology will outpace the technology of downloading."

"We're headed toward digital prohibition."

"We will see two models of Internet use emerging.

> "I don't believe peer-to-peer will proliferate, as security and privacy issues increase the downside for consumers of unstructured and illegal sharing."
>
> —Anonymous respondent

One will be the commercial Internet, where information is paid for. The other is the open-source Internet that is more artistic and chaotic in a positive way. We will come to associate the commercial Internet with 'credible' and the open-source Internet with 'you never know what ya gonna get', though these perceptions will not necessarily align with the actual status."

"Unfortunately, I think the 'wild west' days are over for some forms of anonymous sharing."

"The question is appealing, but I challenge it because I think there will be a splitting of behavior. I think that for much content, such as movies and popular music, ...there will be a shift to online sales and a lessening of the appeal of illicit sharing. I agree that the tools will be available to the vast majority of users, but I think as legal downloading gets easy and if competition drives the price down, this market will grow. But peer-to-peer may be used for other classes of digital information."

"Peer-to-peer networks, while largely focused on illegal activities, are inherently hard to shut down. Recent court rulings have provided the legal cover they need to keep going."

> "The advent of digital-rights-management technologies imbedded in digital products, combined with the rise of identity-based 'authentication' for network access, will make it harder to copy and distribute, but not impossible."
>
> —Anonymous respondent

"The prediction mixes two concepts, 'digital products' and 'free replication.' There are really two kinds of products: Those that are placed deliberately in the public domain, and those over which authors want to retain control. I believe that P2P networks will evolve so that public domain works are easily replicated, while controlled products can only be copied in a controlled manner."

"IP interests will continue to push for defensible limitations on redistribution."

"Much content will still be freely available, but some will not. Workable DRM systems will protect the latter."

"I think these are impossible to stop, and so something must be done to create a flexible digital-rights-management system."

"It's human nature to do this, so yes. Peer-to-peer networks are pretty much the same as tape trading, just easier and with vast availability. We in the music industry have to get off our high horse and make it more desirable to purchase media vs. trading it. iTunes has done a great job, in my opinion, of making the online acquisition of music a good experience for the user, and this needs to be encouraged. In this model, we could actually save money on manufacturing and distribution costs if the volume was comparable."

"There's no stopping P2P."

"Digital file sharers will always stay one step ahead of programmers, regardless of how advanced digital copyright protections become."

"New systems will be designed all the time to circumvent IP safeguards."

"As much as any individual might want to see 'information be free,' the rights of intellectual property holders will not be changed within the next 10 years. Current trends lead me to think that although there will always be renegade systems, the vast majority of individual users will be constrained by technological devices/software from copying and distributing legally protected materials freely."

"Media is the new exchange economy. It is fungible. What is copy? Digital sampling. Media constructivism, is a form of expression. Remixing, sampling, recasting. See redvsblue.com and spooky's new KKK remix. DJVJ, Control Room, we are all producers and consumers of media. We need to get out of our own way. PPV is the model. Open up the channels. Cable, utilities, and telecom are holding us back. The Net's new form is blending realities and structures of physical, virtual, and biological. Web pages and bookmarks are a rearview mirror. Cyberspace has evolved to a new state. And most cannot see it because they are not part of the cognitive tribe. We must pierce the veil of media as artifact and understand it as a form of expression to ever be able to postulate where it will be much less where it is or what form it takes as a verb. As performance art. As expression."

> "Like spam, it's an arms race. The means may change, DRM may change, but file sharing is here to stay."
>
> —Anonymous respondent

"The 'vast majority' of Internet users don't do this now, and won't, even in the next 10 years. They may be able, but the question implies that ability equates to action, which is not the case. And many users don't find current P2P networks at all easy. In the next 10 years the intellectual property owners will devise several means to deliver content people are willing to pay for—to avoid litigation, because it's better than the free version, because it's safer than the free version, because it's easier than the free version, etc. By 2014, we'll see a variety of models ranging from piracy to luxury premium services."

"'Freely' challenges too many entrenched ideas about intellectual property. I hope we find ways of lowering costs and increasing the flexibility of sharing, but digital property does have value, and we need to find ways of appropriately compensating their owners."

"Other legal forms of distribution like podcasting will make this statement appear as dated as the Edsel."

"They will not need peer-to-peer networks to copy and distribute. They will set up their own file sharing, and they will crack copy protection."

"I think—no, I hope—that there is an intellectual-property commons that opens access to a much wider range of information and resources. If it comes to be, then it may not be 'anonymous' peer-to-peer networks. Instead, P2P networks may be visible and supported by the community."

"The culture industry is spending millions of dollars on legal and technological protection for their products and they are trying to destroy free speech, educational use, and first sale. They have the time and might to sue users, one by one, and are doing it. Congress is lax and does not understand the issue."

> "There will be pirate undergrounds, but not used by 'the vast majority.' Laws and prosecution will continue to dampen the activity of some. Copy protection at the software and hardware level will limit the activity for others. This is an area of concern of mine. The government continues to feed the corporate money machines via copyright extensions and heavy-handed restrictions. I fear you'll have to pay just to visit a library."
>
> —Anonymous respondent

"This practice has already been highly curtailed. Media giants will find ways to get their pound of flesh for the use of their products. The question I have is whether independent artists and other creators will find a way to get some kind of return from their work that is distributed via the Net."

"Digital copyright law will adapt to technological and networking advances, and keep the status quo. Some people will copy and distribute; the majority won't."

"Ad-supported products/content and micropayments will eliminate many of the copyright infringement issues not related to music or movies. The entertainment industry will continue to be attacked for its retrograde policies when it comes to licensing noncommercial use of content. Use of cryptography will only serve to alienate customers and drive up the payoff from hacking."

"Peer-to-peer networks will be replaced with inexpensive server farms, and high-speed networks will allow users to connect to content they want without the insecurity of peer-to-peer invasion. The spyware-rich software that enabled peer-to-peer also increased skepticism about their reliability."

"The smart companies will work with this need for free information, not against it."

"Cartels that control the entertainment media will continue their relentless attack on consumer rights, and I expect them to be successful in driving file-sharing activities underground. The majority may still be able to copy and distribute digital products, but not easily and freely."

PART 6

CIVIC ENGAGEMENT

PREDICTION: *Civic involvement will increase substantially in the next 10 years, thanks to ever-growing use of the Internet. That would include membership in groups of all kinds, including professional, social, sports, political, and religious organizations—and perhaps even bowling leagues.*

Experts' Reactions	
Agree	42%
Disagree	29%
Challenge	13%
Did not respond	17%

Note. Since results are based on a nonrandom sample, a margin of error cannot be computed.

In his book, *Bowling Alone: The Collapse and Revival of American Community,* political scientist Robert Putnam argued that one major reason for the decline in civic engagement in the United States is the reluctance among younger people to participate in community groups (e.g., bowling leagues). Other social scientists express concern that Internet use may prompt people to withdraw from social engagement and abandon contact with their local communities. The Pew Internet & American Life Project, however, has documented a vibrant online world where many Americans use e-mail and the Web to intensify their connection to their local community and far-flung groups of people who share their interests.

The debate over whether the Internet is helpful or harmful to civic involvement will continue for many years, judging from the reactions to this prediction. The same percentage of experts disagreed or challenged this statement as agreed with it—and even those who agreed with it often had questions about what it means.

For example, Mark Rovner, an online fundraising expert with the CTSG/Kintera, wrote, "Maybe. But the memberships may be loose and ephemeral, coming together around an issue or a need at one moment, then dissolving and reforming elsewhere the next. Case in point—the Dean campaign."

Many experts scoffed at the notion that a network of computers could increase civic participation. One wrote, "The Internet is a tool, not a replacement for life. The same predictions were made about radio and television when first introduced and those latter day prophets have been forgotten, as have the historical memory of their similar predictions." Another wrote, "The network enables greater civic involvement, but does not spawn it: a desire for change does."

Ken Jarboe, of the Athena Alliance, a Washington, D.C. think tank focusing on the social and economic implications of the Internet, summed up many experts' opinions about how the Internet will not alleviate the time crunch: "The Internet will give people a greater ability to participate—but our limited amount of time (and limited interest) will continue to be barriers to further participation. There are still only 24 hours in the day—and many, many other demands on our time."

Other challenges to the prediction cautioned against exaggerating the Internet's impact. One expert wrote, "Ha! So much of online group 'involvement' is passive. Look at things like listservs. Such a small proportion of members do anything. That is not involvement. Will we be members of more groups? Sure, but that is not social involvement." Peter Levine, deputy director of the Center for Information and Research on Civic Learning and Engagement at the University of Maryland, wrote, "Membership in disciplined, rule-based organizations will probably continue to fall. Membership in face-to-face organizations may also weaken. Counting mailing lists as 'associations' can be a mistake."

> "Even when some activity belongs to just one in a million persons, it means there are 6,000 of those people in the world. The Internet allows those 6,000 people to find one another more quickly and easily."
>
> —Anonymous respondent

However, there are some experts who foresee a rosier outcome. Rose Vines, a freelance technology journalist for *Australian PC User* and the *Sydney Morning Herald*, wrote, "Involvement won't necessarily increase in numbers, but it will in depth and richness of experience. As more groups discover ways to use the Internet to connect, disseminate, and influence, a new element will be added to group interaction. We'll also see those traditionally excluded from participation, including the physically disabled and elderly, being brought on board."

Tobey Dichter, founder of Generations on Line, went even further, writing, "This is one of the most effective uses of the Internet—building community. Exciting global connections will do more for international understanding and intergenerational respect than any tool since the printing press."

ADDITIONAL CREDITED RESPONSES

Many other survey respondents shared incisive, overarching remarks to the question about civic involvement. Among them:

> "What's interesting about civic involvement is that it doesn't have to link to any particular real-space sovereign. I can imagine leagues and guilds emerging very soon in which we'll

be closely involved. Virtual worlds are already pointing in this direction. So our understanding of 'civic' will be much more interesting and diverse than it is now." —**Susan Crawford**, a policy analyst for the Center for Democracy & Technology and a Fellow with the Yale Law School Information Society Project

"The Internet is a great political organizing tool and we will see the interactive, town hall meeting online along with a networked electronic, paperless, voting system, locally and nationally. We will see a return to civic activism. Neighbors will be in a much stronger position to leverage their collective political power quickly and efficiently to make their positions known. Edmund Burke would have a tough time in the digital age." —**Bradford C. Brown**, National Center for Technology and Law

"Without a doubt, the ease to which people can connect to groups online and participate will increase civic involvement…My concern is if it will be primarily for the betterment of the world or simply to assist people's more selfish personal and professional interests. I'm afraid the latter, at least in this country, where entertaining ourselves seems to be the primary focus." —**Jonathan Peizer**, CTO, Open Society Institute

"The dynamics of public participation are changing because of the Internet and not entirely for the better. People will still belong, and believe, but they will increasingly do so from their homes and in smaller groups, where it is more comfortable, less expensive, and 'safer' for people to engage." —**Brian Reich**, Internet strategist, Mindshare Interactive and editor of the political blog www.campaignwebreview.com

"It's true that civic involvement has increased due to the Internet, but there's still a huge number of people who can't afford access in poorer countries. While that might shift somewhat, it's hard to believe that the vast majority of the globe's inhabitants will have Net access or even a computer." —**Mark Glaser**, *Online Journalism Review*, Online Publishers Association

"According to Peter Drucker, volunteerism has been on the rise for many years (predating Internet) and is an important factor in our society. The Internet facilitates volunteerism but the urge to do it does not rely on the Internet." —**Peter Denning**,

Naval Postgraduate School, Monterey, California, columnist, *Communications of the ACM*

"I fear that the Internet is simply rearranging our leisure time and instilling absolutely no additional sense of civic responsibility in people. Web users are fickle, personal, and selfish. They seek the information they want, and act on it, rather than allowing themselves to be swayed or spurred to action by others." —**Mack Reed**, Digital Government Research Center, USC

"I see civic involvement going down as people have more things to do and less time to do it in. A smaller, more radical group of people will dominate civic life." —**Simson L. Garfinkel**, Sandstorm Enterprises, an authority on computer security and columnist for *Technology Review*

"Yes, but their ability to influence public and civic affairs may be counteracted by increasingly 'conservative' legislation."
—**Douglas Rushkoff**, author; New York University Interactive Telecommunications Program

"Certainly, e-rulemaking is a bright spot. NGOs will work to make their, and their members', contributions more useful also." —**Timothy L. Hansen**, MoveOn.org

"I agree that civic involvement will increase, but not in the way that we traditionally measure it—through voting or organizational memberships. I think the Net will create entirely new forms of civic participation, much of it using new media tools. Consider the creation of political advertisements, movies, and Web sites in the current presidential campaign—mostly done by individual. These are new routes to civic engagement." —**Jan Schaffer**, J-Lab: The Institute for Interactive Journalism

> "Increasing civic involvement is not in the interest of traditional sources of political power in the U.S., so the existing power networks are going to put up roadblocks to delay developments along these lines until they put mechanisms into place that will preserve their power."
>
> —Robert Lunn,
> FocalPoint Analytics,
> senior researcher,
> 2004 USC Digital Future Project

"Our definition of civic involvement will also evolve—we will worry less about bowling alone and more about playing video games alone." —**Douglas Levin**, policy analyst, Cable in the Classroom

"I agree with this and see lots of evidence of this already occurring. The Internet is growing rapidly as a powerful social network for connecting with others over space and time. One of the greatest benefits of the Internet is as a social network for interacting with many others and this Net attribute will surely continue to grow as more people gain access and grow more comfortable with electronic communication." —**Gary Kreps**, George Mason University, National Cancer Institute

> "People have a limited amount of time, they cannot keep increasing their involvement in everything. Where is this time to come from? For now, data seem to suggest that those who are already engaged in lots of activities are more likely to then get engaged in more."
>
> —Eszter Hargittai,
> Northwestern University

"People will certainly be introduced to more possibilities through the Internet. But whether that actually translates into more civic involvement is not automatic; it depends on the forces that drive how this process works. There needs to be a real desire to promote social capital that informs the development of these services, and the people running organizations need to understand better how to leverage them." —**Gordon Strause**, Judy's Book, a social-networking Web site

"Many more of our scheduling, occasional access, and connectivity activities will pass through the Net, e.g., signing up for political campaign activities, taking part in online video town meetings (political or social), [and] electronic voting. But such activities may be replacements for current mail, phone/voice, or sign-up sheets. The eventual nature of the network, the price, and the user-friendliness will affect how deeply casual users will plunge into civic applications in ongoing patterns." —**Gary Arlen**, Arlen Communications Inc.

"I'm not so sure about the bowling leagues because I think much of the civic engagement will be virtual. What we now see in the blogosphere will engage more people of different ages, backgrounds, and interests." —**Lois C. Ambash**, Metaforix Inc.

"Civic involvement will increase substantially, and I think the Internet may enable this. Much of the change, however, is highly contingent on economic and political drivers." —**Alex Halavais**, State University of New York at Buffalo

"The Internet will strengthen social and political groups by increasing intragroup communication." —**Stanley Chodorow**, University of California–San Diego, Council on Library and Information Resources

"People are seeking more ways to connect with, share, and take action with like minds. The Internet makes this easier than ever and will accelerate civic engagement in all of its forms." —**Christine Geith**, Michigan State University

"I believe that the ability to register to vote online has already demonstrated the medium's ability to motivate those who can't be prompted to get off the couch. Now they don't have to in order to participate in public discourse." —**Michelle Manafy**, editor, Information Today Inc., *EContent* magazine, *Intranets* newsletter

"Growing use of the Internet does not necessarily mean that people will devote their time in civic involvement. The Internet may provide an opportunity for people to participate online, but it does not 'make' people participate." —**Joo-Young Jung**, University of Tokyo

"Finding people and groups that directly align with your interest will become easier as more people come online, although I am a true believer that the Internet adds and does not replace everything in the physical world. One only has to look to meetup.com to see how the Web site has organized

> "**Civic involvement will see the formation of new kinds of groups, particularly interest groups that cross geographic boundaries. People will have the ability to band in multiple groups based on specific niche issues. Party loyalty will erode as politicians develop platforms made up of these niches, and steps toward a coalition style of government will be realized.**"
>
> —Scott Moore,
> Charles and Helen Schwab Foundation

people with similar interests with real-world gatherings."
—**Tiffany Shlain**, founder, the Webby Awards

"This is likely, but even virtual civic involvement will be 'messy' (loud, sometimes ugly, occasionally enlightening, potentially useful for hearing from a wide range of people) as it is in the mechanisms we have now, such as public hearings. I hope that over the next 10 years effective online tools will be developed that can enable more people to be heard, creating a richer dialog. Or will the growing rudeness and 'me-first' attitudes that we see now—or worse behavior—be what we see in 10 years online?" —**Barbara Smith**, technology officer, Institute of Museum and Library Services

"So we'll be 'bowling alone' on the Internet? We'll use the Internet to support our activities, and technology will facilitate new communications processes not possible in the real world, but the fundamental urge toward civic involvement won't be stimulated by the technology—it has to come from within each individual. Technology might make it easier to register to vote, but you still have to figure out who to vote for, and why." —**Meg Houston Maker**, user experience designer

> "Rather than broadening people's worldview, the Internet easily permits left-handed, D&D-playing, Rush Limbaugh fans to interact only with others who share their worldview. The risk is that civic involvement will decline, as people retreat to the comfort zone of the like-minded. The days of the bowling league where you might meet a smart, committed family man whose politics differ from your own are waning, and the loss of mutual understanding is genuine."
>
> —Perry Hewitt, marketing consultant

"Political scientists and sociologists, while recognizing the high importance of civic activity have, to date, been unable to 'dissect' the mechanisms of civic involvement. Indeed, since the '70s researchers, politicians, and pundits have hailed the role that technology will have in increasing civic engagement, yet this has not happened on a wide scale. Some research even points towards a decrease in

civic engagement and a fragmentation of the public that does not contribute to the whole. I find this statement a bit utopic/deterministic in any case—civic involvement is affected by people, not technology—the technology merely mediates."
—**Michael Dahan**, Ben Gurion University of the Negev, Department of Comparative Media, Israel

"Traditional civic involvement leans too heavily on a 'guilt' model of coercing participation in face-to-face modes, like getting people to bring something for a bake sale. People will do it, and even enjoy doing it, but the participation level is unsustainable. The effect of online communities and connections on civic involvement ought to send us all to Hannah Arendt, on the spontaneous eruptions of civil societies and common feeling, outside modes of power, coercion, and guilt. Online communities and cultures self-select, and because of the more tenuous nature of the connections, are less affected by guilt...So how does anything get done in civic groups based in online communities and connections? That is the fascinating part, and something I've studied in my online ethnography. Things happen because the motivation factor is very high...The righteous guilt and obligatory nudging just isn't persuasive in online communities. If someone doesn't want to be someplace, she just bugs out. The group reaches the end of its life cycle and participants self-select the next thing that captures their attention honestly, instead of out of guilt."
—**Christine Boese**, cyberculture researcher, CNN Headline News

"Civic engagement has not been increasing. College students who are certainly heavily involved in the Internet are not generally interested in civic engagement. The American Democracy Project, in place in nearly 200 higher education institutions throughout the nation, is designed to bring civic engagement to the front burner and produce scholars who also understand the need to give back to their community. If we have to create a structure to promote civic engagement to this strong Internet target audience, then I don't know how the Internet can be responsible for a substantial increase in civic engagement in the future." —**Stephen Schur**, Ramapo College of New Jersey

"Unfortunately, one of the Internet's greatest strengths—creating and bolstering affinity groups—poses a real danger.

The sheer volume of information and potential relationships allows for and even necessitates individuals filtering their content and social contacts. On the contrary, as Robert Putnam detailed in *Bowling Alone*, our society is shifting away from communal activities. The greater the dependence on information technology, the more rapid will be our exodus from community. By its nature, the Internet breeds a certain self-absorption. I think civic malaise will be higher than ever before 10 years from now." —**Daniel Weiss**, Focus on the Family (Christian ministry)

> "The Internet is a tool for social use, not a determinant of social behavior. It does not necessarily follow that making something easier will make it happen. There will have to be significant social changes for this prediction to be true; people will have to have the inclination to use the Internet for this purpose. We have to be careful not to be technologically deterministic—society will determine how the technology is used."
>
> —Susan Kenyon,
> University of the
> West of England, U.K.

"Wishful thinking. My research shows that most people are lurkers both online and FtF [face-to-face]. And the few who are really active online are also really active FtF. It's the TV that's the problem, folks. Once people get off the TV, then they may become more actively engaged in both FtF and online." —**Anita Blanchard**, UNC Charlotte

"The Nets are likely to increase fragmentation rather than an appreciation of collective understanding. So far, the hope that the Internet would remain an aspect of the commons has not proven to be true, despite the best efforts of early adopters. Instead, it has become increasingly proprietary and fragmenting. While we might hold out hope that the decentering impact of the Nets may leave individuals and groups feeling more empowered, it strikes me as pretty unlikely. However, my main challenge to this prediction is that the Internet is not likely to be a key factor in determining civic involvement. Growing cynicism on the part of the intelligentsia, increasing privatization on the part of the economic elites, and rapidly expanding anti-intellectualism, antitechnology sentiment, and

fundamentalist fervor on the part of those who feel excluded by the Nets are likely to be far more significant factors."
—**Alec MacLeod**, California Institute of Integral Studies

"I would say instead, social capital will increase substantially in the next 10 years, thanks to the ever-growing use of the Internet. It will grow through informal online social, business, political, and cause-related networking with the occasional face-to-face meetups." —**Leonard Witt**, PJNet.org

"Civic involvement will expand, but not primarily because of the Internet. Having the ability to communicate and get involved does not impact the desire to get involved. And, while it makes it easier to get involved initially, the Internet does not really dramatically decrease the time needed to be truly involved in an organization. Civic involvement will go up because the Millennial Generation is inclined to be more socially involved. As this generation moves into the workplace, they will take that urge to make a difference and integrate it into their approach to everyday life. That will also motivate those around them to get more involved. So will involvement increase? Yes. Will the Internet play a role? Yes. But does the Internet cause that involvement? No." —**Mike Witherspoon**, Connexxia

"This election is like no other. I remember sitting in a conference in Syracuse, N.Y., when they discussed meet-ups. The house parties and the meetings are the social networks that enable the communities to meet up. At the conference, there was little evidence of minority involvement in the movements, but it took a little time. There was a meet-up at Ben's Chili Bowl in Washington. That is in the heart of a black part of the city.

> "If this happens it won't be because of the Internet (like any other social change considered in this study) but because the citizens will have found the need, will, and readiness for it. And then, yes, the Internet among other things will possibly help."
>
> —M. J. Menou, AmICTad

The book clubs on Oprah, the synergistic listservs, [and] the media groups working worldwide are just a small part of the ways in which citizens meet. I like being asked to be involved in the political process. I have given money, participated, and followed actions that I might not have thought of." —**Bonnie Bracey**, George Lucas Educational Foundation

"Civic involvement will increase, but also the very definition of what constitutes 'civic' will change as well, away from the Putnam-style definition this question uses. Read Schudson for more details, especially his idea of 'monitorial citizen.'" —**Travers Scott**, 9099 Media, University of Washington

"The Internet as a communications medium has clearly made it easier to contact people and organize events, meetings, even recreation. These aspects are valuable to all users of whatever background, and they will continue to be very valuable and lucrative areas for development." —**Mike Weisman**, Seattle attorney, activist in the advocacy groups Reclaim the Media and Computer Professionals for Social Responsibility

> "The increase in civic involvement will come through the weak ties that are the Internet's strengths. This will not necessarily mean greater involvement in local organizations. The pace and complexity of people's everyday lives limits their involvement in groups, and that won't change. They can only deal with a limited number of strong ties. The weak ties of the Internet, however, are available on an 'as-needed' basis. The nature of people's civic engagement will change with the Internet and be different than in the past."
>
> —Kim Keith, About.com

"The only way that civic involvement will increase will be as a backlash against the Internet and technology, not because of it; I suspect that time will come, but not in the next decade. For the next 10 years, like the last 5, we'll be enraptured with and distracted by the new technology. We'll sit in our homes and offices, staring at the screen, and not realize how empty our lives have become. It will take a much bigger social disruption than 9/11 to shake us up, and then some folks will opt out, or greatly scale back. But if you follow the technological innovation from telegraph to radio to TV to the Internet, it's hard to believe that folks will choose demanding forms of social interaction when the Internet as entertainment-delivery service is right there in their houses, throwing big screen images of fun on their walls without ever having to leave the house." —**Peter Eckart**, director of management information systems, Hull House Association

"There will be some increase, and more social mobilization, but institutional power will remain resistant to change."
—**Barry Wellman**, University of Toronto

ANONYMOUS COMMENTS

The following excellent contributions to this discussion about the Internet and civic involvement are from predictors who chose to remain anonymous.

"Certainly, the Internet is being used more and more for organizing, including the civic organizations mentioned. I just don't want to overestimate this, as more information has not made the U.S. more involved in politics in the past 20 years. Still, the meet-up type activity will increase substantially."

"I had a conversation with Robert Putnam (*Bowling Alone* author) about this exact proposition. The city of Seattle's civic involvement initiative, known as the Democracy Portal, is based to some degree on this premise; we do agree that community health is related to participation, but believe that we can use the Internet to help promote it. Although I checked 'Agree,' this is a really hard one to make happen, and I'm certainly not sure it will. But we are trying."

> "It seems to me that there is less civic involvement of a substantive nature as people become dissociated with their local communities and shift their attentions to virtual communities that come and go, often at the whim of the larger media conglomerates."
>
> —Anonymous respondent

"I don't anticipate a big up-tick in such activity, nor do I expect a big change in the incidence of bowling alone in America."

"The Internet will enable more participation of a less committed variety, but it may not have the same desirable consequences of participation when participation was harder."

"Too many factors affect civic involvement for the Internet to be perceived as the prime mover. Yes, information-intensive activities, such as elections, may reach more people via the Internet, but will it make nonvoters go out and vote? I doubt it."

"Civic participation will grow but not because of the Internet. The Internet will help facilitate such groups and perhaps attract some people to them, but it will not be the 'prime mover' of increasing civic participation."

"The Internet will not be magic pixie dust that causes everyone to reach out into their community. I believe it is more likely that the involved will become more so. Soccer people will become more knowledgeable and networked through the Internet. I don't believe the Internet will awaken social and civic consciousness in the masses."

"Technology is a tool. A book can connect us to the world, or help us escape it. Same with the Internet. There are no guarantees. People do not obey Moore's law."

> "Virtual interaction will replace physical interaction. Since the former creates less close ties than the latter, the overall sense of connectedness will actually decrease."
>
> —Anonymous respondent

"The surveys and trial projects on this are very positive, but the actual results are pretty negative. Everyone says that they will participate, but few do—except in small, closed user groups such as law faculties."

"Young people in particular are making enormous use of tools like Friendster to expand social networks."

"People who do it offline will do it online; people who don't join groups just don't, no matter which environment we're talking about; the new generation of Internet users—today's college kids and high school kids—feel differently about online community than earlier generations, maybe because it has always been around and they have been cautioned about the 'Internet stranger' from a young age on. Most do not even believe that one can form close relationships online."

"Civic involvement does not need the Internet, it needs a lack of apathy. Sitting in front of a computer surfing the Web is likely only to disengage individuals from their more proximate surroundings, not engage them."

> "Civic involvement means deliberative discourse, human engagement, understanding all of which requires being informed and engaged. The Internet is a tool, not a replacement for life. The same predictions were made about radio and television when first introduced, and those latter-day prophets have been forgotten, as have the historical memories of their similar predictions."
>
> —Anonymous respondent

"I believe the Internet will be useful merely as a tool to support social networks."

"Civic engagement is a function of time and attention. Neither are replicable resources, and there is nothing in the Internet that provides more of either."

"Human ability to interact and find time is limited; electronic communities amplify this ability but not by an enormous factor."

"People only have so much time. There are many opportunities for involvement now. More opportunities won't increase civic involvement, although it may alter it somewhat."

"I do see this happening, e.g., use of e-mail lists by homeowner associations, but I don't see it leading to substantial civic involvement. Civic involvement implies engaging in a dialog about ideas. Much of the use of the Internet seems to be between like-minded individuals reinforcing shared views. Where's the dialog?"

"America has always been a pluralistic society. The Internet merely provides a vehicle for those inclined to accelerate that pluralism. However, civic involvement will not increase as a percentage of the population."

"The trend continues to be that groups are adopting the Internet to help them organize as soon as a critical mass of their members have the capability to participate."

"Involvement in online groups may, in fact, erode traditional civic participation."

"Yes—but as campaign 2004 is suggesting, this kind of civic involvement has both positive and negative aspects. In the first half of this year, the positive aspects were clear as groups like movon.org were getting people more engaged with the political process, increasing voter registration, etc. But as we've lived through the various blogger wars surrounding this campaign, not to mention the various Truth campaigns (Texans, Swift Boat Captains), it is less and less clear to me that the effects are purely positive. The negative campaign tactics are being played out on the grassroots level and they are resulting in more antagonism and divisiveness than I would have imagined possible. We will have lots of work to do to repair the damage to civic life at the end of this process."

"Special-interest groups have become much more effective in applying pressure and raising funds due to the Internet. This has even more deeply woven together money and politics, making it even more difficult to make progress on campaign finance reform. The result will be increasing cynicism and decreased civic involvement long term."

"I think it will decrease. However, contributions to organizations probably will increase as Internet solicitation and the ability to transact easily intersects with ubiquity and awareness to create a less intrusive/anonymous way to offer one's own support with time/money without getting involved."

"It depends on how you define 'civic engagement,' including the definitions of membership, but I suspect distraction and withdrawal from meaningful links in favor of thinner ties is likely."

"This will happen if people are motivated—if they are 'mad as hell and not going to take it anymore.' The network enables greater civic involvement, but does not spawn it: a desire for change does."

"The Internet will have a negligible effect on total civic involvement, though it will alter means."

"Initial studies have shown that despite the Internet's claim to be connective and community building,...the Internet can leave individuals feeling isolated. Like television, the act of

Civic Engagement

staring at a computer screen will not increase civic involvement unless there are personal incentives that encourage it."

"Will more people be connecting with one another and joining online communities? Yes. Is this traditional civil society? Not sure. Many of the groups could create [a] honeycomb structure of walled-off little communities that don't meet in any kind of civic public space, virtual or physical."

"I think this prediction underestimates the basic human wish for human company. Time and again, people—including experienced Internet users—articulate the notion that while Internet communications offers all kinds of useful networking possibilities, they still choose real-time face-to-face interaction, group activity, and communication to fulfill basic human social needs."

> "People's capacity of social interaction is saturated even without the Internet. The limit on the number of social groups one can be a part of is not the size of one's mailbox, as it were, but the time it takes to communicate with each group. The last time I checked, there were still 24 hours in a day."
>
> —Anonymous respondent

"Depends, of course, on how you define 'civic involvement.' The current presidential campaign has increased involvement. Will this last? Will this be truly good for society, or is this just a case of divergent groups becoming more well defined, more isolated, and shouting at each other? Will it have the impact of league bowling? I fear not."

"It's not obvious how the Internet will increase face-to-face social interaction. So many other variables are at work in determining how much time we spend in nonwork related activities, including: (a) how many people need to work overtime, two jobs, or all the time; (b) the independent religious beliefs of the populations (some groups attend church more than others); and (c) the growth of cell phones as an alternative to computer-based use of the Internet."

"I think that in the next 10 years, we will need to think more about what it means to participate in all areas of our life. What does it mean to participate in a virtual protest, for example? Does this constitute civic engagement? Perhaps, our definitions of civic engagement are changing to take into account new digital forms of participation. We need to think about and study the implications of these shifts."

"Pew studies found that in comparing Net users and nonusers, Net users do more sports and recreation. Why? My theory is because organization and administration of sports clubs and teams is generally administratively difficult, and the Internet has fantastically facilitated small community organization. People can spend more time playing and less time on the phone, organizing playtime. There is less of a barrier to playtime. I also think that features like RSS will improve frequency to noise ratios—people will be able to narrow in on their interests and be empowered, instead of suffering from information overload."

> "Already we see dart leagues, kids' soccer, and other not-for-profits organizing on the Net. What groups don't? Traditional men's dinner clubs like Rotary. They are already in a coma and will go the way of felt hats."
>
> —Anonymous respondent

"It would be great if this would happen, but all indications I can see point toward more privatisation/individualisation rather than collectivisation. Civic involvement at the level of 'citizenship,' or involvement in governance, is hampered by a growing sense of alienation from political processes and cynicism about what difference an ordinary person can make. Modes of membership in social and leisure groups will be affected by Internet use, and it's likely that people will be more easily able to find like-minded people who share their interests."

> "People do not obey Moore's law."[1]
>
> —Anonymous respondent

"While the Internet facilitates the growth of vast interest-based networks, it unfortunately accelerates the decline of genuine civic involvement. Local life is being eclipsed by virtual life. Local identity is being eclipsed by identification with groups that have nothing to do with geography. Civic life begins at home, but the concept of home is being undermined."

"Heavy Internet users are way too distracted and multitasked to make commitments to groups like this for a long period of time. Most Americans are too selfish and self-involved to commit to anything more that gets in the way of their convenience-driven lifestyles, unless they're already passionate about a cause (like a religion, a sports team, or a political candidate) and can be whipped into a frenzy by marketers and demagogues."

"The Internet certainly does foster civic/group involvement. It is easier to be part of a community that includes contact via the Web. For one thing, you can check into or sign up for things going on while at work. A second thing is: You can schedule group events easily (e-mailing bunches of people at once). Third, I've been able to download all kinds of paperwork connected to civic and group involvement, like newsletters from churches and neighborhood groups, voting registration forms, etc. Finally, it is much easier to investigate or find out about groups in the first place via the Internet also. The Internet allows one to travel far and wide while sitting at a desk. But the Internet won't replace face-to-face group activities."

> "It's happening now, from neighborhood kids' soccer clubs having Web sites to nearly any type of organization. It's the way of the world. Everyone has something to say and to share. That's human nature and the Internet has allowed us to have the widest possible audience. Yes, there will be even more than we can currently fathom."
>
> —Anonymous respondent

"People's willingness to use information merely to confirm their beliefs, irrespective of the relevance of the information in question, will win out in the majority of cases over the ability of the Internet to expose people to unprecedented amounts of information and breadth of perspective. I believe the main effect the Internet will have in these areas is to increase the convenience, or perception of same, of business as usual. E.g., it is not an 'increase in civic involvement' when 100,000 get the letter-to-the-editor boilerplate Astroturf text online and e-mail it to their local papers; it's just a technological enhancement of the marketing of ideas. Also, a 'substantial' increase occludes the notion of the digital divide, which is still alive and well. By and large, Internet resources increase the convenience and sometimes power of endeavors that were already within reach for the individuals in question."

"At the end of the day, involvement is limited by time. I see selective involvement still, with greater involvement being channeled into areas where there is greater payoff. You could say that involvement will become deeper instead of broader. This could mean that strong civic groups become stronger while already marginal ones shrink and die."

ENDNOTE

1. In 1965 Gordon E. Moore, cofounder of Intel, accurately predicted that the capacity of computer chips would double every two years. See http://www.intel.com/research/silicon/mooreslaw.htm.

PART 7

EMBEDDED NETWORKS

PREDICTION: *As computing devices become embedded in everything from clothes to appliances to cars to phones, these networked devices will allow greater surveillance by governments and businesses. By 2014, there will be increasing numbers of arrests based on this kind of surveillance by democratic governments, as well as by authoritarian regimes.*

Experts' Reactions	
Agree	59%
Disagree	15%
Challenge	8%
Did not respond	17%

Note. Since results are based on a nonrandom sample, a margin of error cannot be computed.

There was hardly any disagreement with the assumption underlying this question: More devices, appliances, and other objects would be tied to the information infrastructure. Proponents of automatic identification devices say convenience is one of the best outcomes of embedded technologies, such as electronic toll collection, mass-transit fare cards, and building-access key cards. Our prediction took a different tack, eliciting fear, defiance, and hope that it might be prevented from most of the respondents.

One expert seemed to welcome the surveillance, writing, "In response to terrorism, such surveillance is necessary and predictable. Its expansion to address criminal activities is also obvious. In many cases, such as tagging pedophiles, it will bring security to the innocent, and antisocial behavior can be controlled. In democracies, why worry?"

Another expert would counter that such "social surveillance" is potentially more harmful than law enforcement: "It goes far beyond arresting people. This will be a method of social control in more subtle ways, too. The risk of being seen as 'different' will grow, and children will grow up with the knowledge that their every move is being watched. This is a recipe for killing the kind of independent thinking that creates innovation, vibrant political debate, and a free society in general."

> "This is likely not just because of 'embedded devices,' but because of eroding anonymity and great pressures to reduce anonymous activity to a negligible presence in society."
>
> —Gary Chapman,
> LBJ School of Public Affairs,
> University of Texas at Austin

Susan Crawford, a policy analyst for the Center for Democracy & Technology and a Fellow with the Yale Law School Information Society Project, foresees a different consequence: "It seems to me that most of this surveillance will be private in nature, and that private firms will be unwilling to make their databases widely available. I agree there will be lots of surveillance, but I don't see it being turned over to government authorities. Instead, it will be used to market to us in ever-more-personalized ways."

But, as in the peer-to-peer debate, there is a group of experts who believe in the power of technology to elude the would-be watchers. One wrote, "The innovators of these tools have consistently been ahead of government efforts to counter their influence. This will continue

Embedded Networks 105

to be so, and citizens will continue to 'get away with' activities using technology that government does not understand."

Experts challenged the wording of the prediction, pointing out that it was impossible to measure. One wrote, "Since the earth's population grows every year, 'increasing numbers of arrests' are inevitable. That more powerful devices will offer greater surveillance also is a given. The question is whether government will spy on us proportionately more than they do now, which, by the way, we can't measure because it's done secretly."

Some experts believe the prediction will come true in less than 10 years. One pointed out, "Right now, almost no one knows what RFID means. In 5 years, everyone will." (RFID stands for "radio frequency identification." RFID tags can be as small as an adhesive sticker and can be used to identify people or objects automatically by transmitting a unique serial number.)[1]

The final word goes to the expert who anonymously wrote, "I would elaborate, but someone might be watching."

ADDITIONAL CREDITED RESPONSES

Many other survey respondents shared incisive, overarching remarks to the embedded networks question. Among them:

> "If surveillance tools really get that good and ubiquitous, this should lead to less crime and fewer arrests. (Perhaps a spike up in the interim.)" —**Benjamin M. Compaine**, a communications policy expert and consultant for the MIT Program on Internet and Telecoms Convergence

> "I believe it will take longer to completely roll out, but that we are surely headed in that direction." —**Moira K. Gunn**, host, *Tech Nation*

> "Yes, but there will also be increasing black-market activity." —**Douglas Rushkoff**, author; New York University Interactive Telecommunications Program

"This type of monitoring has been increasing on a regular basis for some time now. This is part of the mission for the NSA. It's also something that can be automated, so it's a natural consequence of the technology. Technology is the great enabler of freedom or tyranny. It's the responsibility of the people to nurture the former rather than the latter." —**Robert Lunn**, FocalPoint Analytics, senior researcher, 2004 USC Digital Future Project

"This is inevitable and worrisome. At some point, probably beyond 2014, the courts, at least in this country, may try to control the use of Internet devices by law enforcement by barring evidence gathered in certain ways from being used in court. But, that process will be very difficult and take a long time to evolve." —**Stanley Chodorow**, University of California–San Diego, Council on Library and Information Resources

> "We will never allow America to become a full police state. Moreover, as technology gets more sophisticated and law enforcement seeks to use it to clamp down tighter, society will find a way around those limitations—crime will still exist, and it will evolve along with the technology. That is innovation at its finest."
>
> —Brian Reich,
> Internet strategist
> for Mindshare Interactive
> and editor of the political blog
> campaignwebreview.com

"The USA Patriot Act has already diminished our civil liberties and made it possible to disguise flimsy excuses for 'probable cause' as legitimate reasons for surveillance. The courts have consistently ruled in favor of businesses' rights to monitor their employees' communications. The prevalence of technologies like GPS and RFID will make it almost impossible to control the extent to which commercial enterprises monitor the activities of customers and potential customers. And so long as we treat the war on terror as an ongoing state of war in the traditional sense, it will be almost impossible to challenge such intrusions." —**Lois C. Ambash**, Metaforix Inc.

"Legal challenges will arise, proceed to the Supreme Court, and be upheld on constitutional grounds." —**Mack Reed**, Digital Government Research Center, USC

"Technologies like RFID put the Orwellian vision of Big Brother an abuse away. We will have cameras watching all our cities, RFID tags in all our products, smart products in our homes—all networked and all the related information stored in a database. As a country we have to be careful not to let our fears and our good intentions turn our networks into *The Matrix* or a revival of the Total Information Awareness Program. All the information will be out there. Privacy will no longer exist per se. It will only exist in the artificial sense in that it will have to be regulated. Our ability to do that will be critical to keeping our own democratic construct together."
—**Bradford C. Brown**, National Center for Technology and Law

"There will be greater surveillance, probably; greater arrests, maybe. But this is a chilling prospect overall." —**J. Scott Marcus**, Federal Communications Commission

"This statement seems to focus on some of the negative aspects of greater connectivity and availability of information. Increases in connectivity will provide us with greater and easier access to products and services, as well as increased surveillance. I suspect a tradeoff of greater access, maybe a loss of personal privacies."
—**Gary Kreps**, George Mason University, National Cancer Institute

"I think that significant limits will be placed on government use of this information—at least in democratic countries."
—**Jonathan Band**, an attorney specializing in e-commerce and intellectual property with Morrison and Foerster LLP in Washington

> "This is happening now. The current terrorist context and the networked nature of the Internet facilitate better surveillance. What occurs politically in the next 10 years will determine if the situation gets better, worse, or if there is a political backlash. If terrorism continues unabated the situation will only get worse because it will give legislators an excuse for laws like the Patriot Act."
>
> —Jonathan Peizer, CTO, Open Society Institute

"Anyone who believes such things is not thinking through the consequences. Possibilities for surveillance are already well

beyond our capacity to keep up with them, as the inability of our intelligence to keep up with Internet 'chatter' now demonstrates. And arresting people doesn't resolve but only begins a process: With courts and prisons overloaded, you have to have much greater faith in the possibilities of a police state than I have to imagine significant increases in arrests."
—**George Otte**, technology expert

"Sadly, this is possible. However, it would be technological determinism to say that this will happen. Why is the Internet inherently good or bad? Surely as Castells suggests what is key is the nature of people. If society becomes less tolerant and more authoritarian then the Internet will assist this. If society becomes more open, tolerant, and participative, then technology will enhance this. Technology is the tool of people, it does not automatically lead them to one route or another." —**Nigel Jackson**, Bournemouth University, U.K.

"Like any technology, uses of embedded computing devices have good and bad implications. I expect that important arrests will occur that will increase the societal value of the technology, and I also expect that the technology will be abused to some degree by all governments." —**Terry Pittman**, America Online, broadband division

> **"Between RFID, sensory networks, Homeland Security, and the Patriot Act, how could anyone bet against an increasingly intrusive surveillance state?"**
>
> —Howard Rheingold,
> Internet sociologist,
> writer, speaker

"What happens in the U.S., I fear, will be much different from that of authoritarian regimes. Still, the effect of the global information flow is, I believe, toward democratization and the institutional arrangements that protect both free speech and privacy. People will want to be in contact with each other and not be subject to observation and arrest. They, therefore, will insist on regulations and laws to protect these." —**William B. Pickett**, Rose-Hulman Institute of Technology

"First, embedded networks are going to take some time to grow to the point where they could provide this kind of infor-

mation. Second, democratic societies will likely see some controls on the use of these kinds of surveillance methods."
—**Ezra Miller**, Ibex Consulting, Ottawa, Canada

"The interests to build in identification of everything from CPUs to clothing using a wide variety of markers—electronic and otherwise—will outpace the ability or interest of the populace to block or thwart these systems. Attempts by constitutionalists, libertarians, the privacy-minded, and individualists will lag behind commercial and security interests in their ability to enact policy to protect against this increased surveillance." —**Dan Ness**, MetaFacts

> "I agree with the statement but would like to add that the public will have greater access to devices and hacks that block or scramble such surveillance devices."
>
> —Bornali Halder, World Development Movement

"Technology has still not proven how it can make a class more efficient than a physical class. It may happen someday, and technology will definitely contribute to what is described in the prediction, but not (on a massive scale) in 10 years and not in the form envisaged. I believe individuals will move more than ever." —**Daniel Kaplan**, France's Next-Generation Internet Foundation (FING)

"The civil libertarians will still be strong in the next 10 years. The real danger of this technology is in 30 years, when there are a couple of generations who have grown up with this and don't see it as an infringement of their rights, but as a legitimate governmental and workplace security tool." —**Peter Eckart**, director of management information systems, Hull House Association

"These kind of paranoid fantasies just make our society less inclined to adopt important new technologies. While other nations, like the Japan, are racing ahead with RFID and other networked applications, we are responding to zealots who want to prevent technological innovation. Besides the technology generally not enabling this kind of surveillance, if there were any abuses, laws would quickly be passed prohibiting them." —**Rob Atkinson**, Progressive Policy Institute (previously project director at the Congressional Office of Technology Assessment)

"Data collection and the ability or desire to process it will thwart large-scale social control. This will also be affected by counter-surveillance and countermeasures by those who really pose a threat." —**Ted M. Coopman**, University of Washington

"There will be an increased ability to track people's movements and activities, either as a surveillance action in real time or through a record check after the fact during an investigation, such as through tracking of cell-phones, cars, and other wireless Internet-connected devices. As the benefits to crime reduction rise, there will be increased tension. Will our traditional ideas of civil liberties stand up?" —**William Stewart**, LivingInternet.com

"Embedded network technologies will become a useful tool for law enforcement. In 5 to 10 years as tracking technology use becomes more sophisticated and widespread, a deterrent effect will come into play decreasing a huge variety of crimes from kidnapping to theft to murder among others. Over time, as the deterrent becomes more real, arrests will decrease in democratic societies but will increase in authoritarian regimes." —**Richard W. DeVries Jr.**, DeVries Strategic Services, St. Charles, Illinois

"This is a *Star Trek* meets *Big Brother* prediction. The integration of computing into our social fabric will be gradual, and people will chafe against infringement on personal liberties. There will be a balance between more information and personal privacy. This prediction also doesn't acknowledge that if everything was wired there would be an immense challenge to dealing with the huge 'infoglut' that would occur." —**Lyle Kantrovich**, usability design expert, Cargill, blogger

"The devices will be every day more independent, and the democracy will increase, too." —**Joao Sartori**, blogger

ANONYMOUS COMMENTS

The following excellent contributions to this discussion of embedded networks are from predictors who chose to remain anonymous.

"We are ill-prepared legally and morally for omnipresent sensors."

"More likely that this trend will chill deviant behaviors (benign and less benign), rather than result in more arrests. Individuals, too, will be surveillors, as well as surveilled."

"I don't think it will be like in *Minority Report*, but civil liberties are at risk if we are not more careful."

"This is a double-edged sword. Digital literacy will expand to include protection of self from retailers, as well as governments."

"Only by constant vigilance will this kind of surveillance be kept under control."

"There will be arrests, and these will be the proving grounds for protecting privacy."

"I think there will be more surveillance, but if the recent reports on the capacity of the FBI and CIA are any indication, law enforcement does not seem to have the capacity to use the increased surveillance. That will probably come over time, but the applications and training take a long time."

"I agree with the possibility, but am not so sure that the surveillance mechanisms for dealing with the massive data collected by these devices will keep up with the growth. Without new approaches to extracting potential patterns, it will be difficult to pinpoint possible threats—there may be more arrests, but they may not be the right ones."

> "While I doubt things will look like *Bladerunner*, I do think since crime will become more ethereal, the way it is tracked will have to catch up. I think you will see a dramatic increase in nonviolent e-crime...it is less messy, easier to cover your tracks, and will be attractive to nontraditional criminals."
>
> —Anonymous respondent

"The pervasive use of technology without well-thought-through privacy and civil liberty protections threatens to change our society in this regard. All one has to do is combine pervasive computing and the full, most liberal interpretation of the Patriot Act, and we make George Orwell's version of *1984* look Libertarian."

"This is a legal rather than technical issue, related to the various terror laws. I do not think the technical capacity for spying will be higher, only that the laws will be relaxed still further and agencies more efficient in cross-matching data (à la David Lyon, Mark Poster, etc.)."

"Certainly more so in authoritarian regimes, but not necessarily more so in democratic governments. However, the use of those devices may make some democratic governments more authoritarian."

"Phones may take on more, but this other stuff just can't happen within a decade. Note Carver Mead's 11-year rule: It takes 11 years from the time we have a credible lab demo. I haven't seen such a demo."

> "It is not a given that embedded computing should translate into control. Embedded devices can be anonymized, and we should expect a greater effort in the democratic world to take advantage of these technologies while we minimize their capacity of tracking people."
>
> —Anonymous respondent

"Greater surveillance possible—yes—but more arrests? Hopefully more privacy restrictions, so information gleaned from any potential surveillance cannot be used, and will be discouraged."

"As the devices proliferate, I have a hard time seeing that law enforcement and government will keep pace with noncriminal use of IT. They will have to ask what will they gain by running continuous surveillance based on these devices. I can imagine these might be resources that could be used, but it's a sci-fi scenario that would have them commonly monitored."

"The private sector and government have increasingly en-dorsed surveillance technologies. Monitoring devices are now pervasive in the workplace, but in the next decade surveillance will move into the home. Insurance companies are proposing instrumentation of motor vehicles; the department of corrections is embracing 'house-arrest' technology; government is moving to pervasive monitoring of public spaces for counterterrorism."

"Unfortunately, if you care about civil liberties, this will likely be the case."

"The number will increase, but watchdog groups will keep us from entering a 'Big Brother' dystopia."

> "Help! This is the demon dark side of the technology evolution—and will be the beast that destroys the marvel of the Internet."
>
> —Anonymous respondent

"We do have watchdogs in the U.S. to make sure our rights are protected. The same cannot be said about authoritarian regimes. Seems very Orwellian, doesn't it?"

"Engineers need to put out RFCs [Request for Comments list messages—shared documents tied to internetworking and network engineering] that would make it very difficult to do this on an unauthorized basis (with no legal warrant). Current Internet protocols make it too easy for governments and corporations with the right expertise to invade privacy."

"The use by government will become entrenched before the lawmakers can do much about if, even if they want to. But in a society where there are no secrets, the value of secrecy declines."

"I hope so, except for the part about authoritarian regimes, but there will be a declining number of those."

"We need policy constraints to ensure privacy. This is a big issue for the next decade."

"We are already seeing consumer backlash on adware and spyware."

"The impact will be small. Mostly 'accidental' discoveries analagous to the occasional video recording of crimes in progress. Criminals will turn the things off."

"I am one of those privacy nuts who believe that we are well along the way toward this kind of ubiquitous monitoring/surveillance."

"Look at the growing use of electronic monitoring and ticketing for traffic violations. We're only just starting to see the use of electronic monitoring by governments. This will become a very serious issue once the public begins to see the full measure of what could happen."

"The next decade will require methods of preventing terrorism. While civil liberties must be protected, the trade off with personal and communal security may result in compromises."

"It seems unavoidable. How many arrests have been made due to credit card records that in an earlier age of cash only would have not been possible?"

"I agree with this assertion, although civil liberties will be vigorously fought for in democratic societies that will slow this trend in those societies."

"RFID, UWB [ultra-wideband signals], great technologies with potentially nasty applications."

"Scandal will come from democracies, but danger from rogue states."

"I am concerned that the civil liberties and privacy issues are not getting the discussion and the creative thinking for solutions that they deserve."

"I am concerned about what happens when ubiquitous computing meets a terrorism-obsessed world. Governments will have the ability and excuse to curtail privacy and civil rights if they are able to continue to capitalize on fear of terrorism. I think that the citizens of many countries are going to struggle with this in the next decade or so."

> "There is no stopping increased surveillance. The climate of fear, which I expect to continue, will only increase the political viability of surveillance as a means of control. Other factors pushing this trend will be the spin-out of cheap, miniaturized surveillance technology from military applications, and increased fear(mongering) due to the aging of the population and increased immigration."
>
> —Anonymous respondent

"As we are further distanced from literate values such as privacy, we will cease to see privacy as a value that ought to be preserved. Not sure if democratic governments will be able to be said to still exist."

"The volume of information will likely be too great for much effect. Possibly review of data after crimes."

"The use of technological surveillance is continually increasing and becoming more sophisticated. Governments and business harness the possibilities that technology offers to increase their information collection and their control. The fostering of fear is simply one measure that will be used to legitimate this information collection (as seen in the U.S.'s legitimization of increased surveillance powers following September 11, 2001)."

"Add to this: citizens doing the surveillance of each other!"

"Unfortunately, I must agree. The imposition of the USA Patriot Act and the actions of the U.S. and other governments against persons perceived to be threatening to the country have hardened determination to invade privacy. Despite aggressive actions by privacy advocates, the press of technology developments will outweigh their ability to curb their more widespread use."

> "This is not really the question. The question is: Will due process and constitutional law keep pace and maintain protections[?] 'Who watches those who watch us?'"
>
> —Anonymous respondent

"Pretty much right on. There may be some pushback from civil libertarians, and by 2014 some sort of legal beachhead may be established in Congress as a result of widespread civil rights abuses."

"This will be a part of the transition of democratic governments into authoritarian regimes. Already occurring."

"No doubt, the potential for the use of embedded computing devices to erode privacy rights will continue to increase. I do think that most Americans and others in historically democratic nations so value the right to privacy and are already concerned about its erosion that a tip of the scale towards less privacy will produce a backlash. That backlash will lead to a whole new set of regulations and laws with respect to the use of embedded devices for surveillance in most democratic countries. The potential for invasion of privacy where no historical imperative exists for the right of privacy will make the use of imbedded technology for surveillance not only likely, but a given."

"It will happen slowly and subtly. Only in cases that are large and well hyped in the media will it be challenged. This is most likely going to happen with earlier cases, eventually surveillance will become an old news story and will not be deemed newsworthy, and then people won't know about it."

> "In a society worshipping at the altar of convenience, who will stop it from occurring?"
>
> —Anonymous respondent

"It's pretty much inevitable that if embedded technologies collect information about us, then that information will be used by others in ways that invade privacy and introduce opportunities for new crimes against the person. I'm less sure about the use of this surveillance on the part of governments (democratic ones anyway) than on the part of commercial operations."

"Sadly, this is likely primarily because of the rationale provided by overreaction to the 'terrorist threat' in the U.S. and the skillful use of that rhetoric by even more authoritarian governments."

"We are already seeing the beginnings of this trend. If there is another terrorist attack, pressure on government to increase surveillance will also increase. Surveillance by business may not be as easy. A consumer revolt is brewing against spyware and the use of cookies to monitor usage."

"I believe certain watchdog groups and concerned citizens will prevent this from happening. As evidence of the prediction becomes 'news,' these groups will fight the application of 'tracking' technology into their personal lives. As long as government doesn't outlaw the choice of 'nonuse' of these devices, I believe the majority of self-determining individuals and manufacturers will be successful on the 'anonymous' side of production/lifestyle."

"This is sure to happen, and it presents an important challenge to all of us who believe in individual liberty. We must think through the way technology changes what is private and develop new concepts of reasonable privacy that preserve liberty and are workable in a networked world."

ENDNOTE

1. For more information on RFID, see http://en.wikipedia.org/wiki/RFID.

Part 8

Formal Education

PREDICTION: *Enabled by information technologies, the pace of learning in the next decade will increasingly be set by student choices. In 10 years, most students will spend at least part of their "school days" in virtual classes, grouped online with others who share their interests, mastery, and skills.*

Experts' Reactions	
Agree	57%
Disagree	18%
Challenge	9%
Did not respond	17%

Note. Since results are based on a nonrandom sample, a margin of error cannot be computed.

There has been an aggressive national campaign to bring computers and the Internet into schools—nearly all American public schools have some kind of Internet access for students. However, schools with the highest concentrations of poverty are the least likely to have

Internet access—and access is only part of the story. Internet-savvy students are remarkably different from their nonwired peers when it comes to tackling homework, communicating with their classmates and teachers, and engaging with the outside world.

Experts cautioned that each new technology—motion pictures, radio, television, and now the Internet—rekindles the hope that it will transform education for the better. Indeed, few experts agreed wholeheartedly with the prediction. The following response sums up that type of reaction: "This will transform learning as we know it. Alas, we may escape the strangulation of the agricultural calendar." More typical was the expert who wrote, "I agree this will happen, but only in the wealthier school districts and private schools. The numerous school systems that can't afford enough computers or adequate Internet access will fall farther and farther behind."

Many of the respondents who have had experience with teaching online said only highly motivated, mature students exhibit the ability to be successful in a learning environment in which so much responsibility is placed upon a student. Moira K. Gunn, host of public broadcasting's *Tech Nation*, wrote, "I do not now, and have never, witnessed successful benefits in virtual classrooms. While the role of the teacher will change from authority figure with all the information to one-on-one educational coach, the one-teacher-one-student paradigm will remain the most effective." Indeed, children in elementary school "still need a watchful eye and human attention," according to one expert.

> "Some students will do this, students in affluent circumstances and enlightened schools. But 'most' students are not so lucky."
>
> —Anonymous respondent

Older students—those in college, graduate school, or of nontraditional school ages—will benefit the most according to many experts, but "establishment" institutions will be the hardest to convince. Gary Bachula, a technology development leader, most recently at Internet2, wrote, "Schools and colleges are enormously resistant to this kind of change—more so than I would have predicted 10 years ago. As a result, traditional methods of learning will slowly start to compete with the 'upstarts,' first, the 'proprietary colleges.' Then, one or more of the older institutions will get aggressive in this arena—and then an avalanche will occur."

Another respondent wrote, "Harvard and other major universities are not likely to go virtual. In fact, being on campus will become a thing of status. Online learning and virtual learning will allow more individuals to go to college, but it will take decades if not centuries for online learning to gain the same status as classroom learning. I see enhanced classrooms and dorm rooms…but not a radical change in how learning occurs."

> "If people are taught to hang out only with those of like interests, mastery, and skills, they will become less tolerant of diversity. More medieval."
>
> —Peter Denning,
> Naval Postgraduate School, Monterey, California, columnist, Communications of the ACM

ADDITIONAL RESPONSES

Many other survey respondents shared incisive, overarching remarks to the question about the Internet and education. Among them:

> "I've spent enough time worrying about distance education to despair of this goal being met. Schools are awfully hidebound institutions. So, although I'd like to think this prediction will come true, I'm thinking the time scale is much longer—perhaps 50 years rather than 10." —**Susan Crawford**, a policy analyst for the Center for Democracy & Technology and a Fellow with the Yale Law School Information Society Project

> "Schools have already lost major share in the market for education. Virtual classes are already happening—we'd better improve their content." —**Bob Metcalfe**, Polaris Venture Partners, inventor of Ethernet and founder of 3Com

> "I agree, but am saddened to think that human interaction will be decreased in this part of the growth and development process." —**Bill Booher**, Council on Competitiveness

"There will be a continued need for structured education."
—**Barry Wellman**, University of Toronto

"This is already happening, where online education is being integrated with traditional educational programs. I like the combination of educational approaches better than pitting in-person and distance education against one another. I think that electronic education is a wonderful supplement to more traditional educational approaches." —**Gary Kreps**, George Mason University, National Cancer Institute

"Students already are doing much research and reading online, and this naturally will continue and accelerate with greater availability of computers and fast networks. But students and teachers always will want some direct interaction."—**Mike Botein**, Media Center, New York Law School

"Not only will students spend at least part of their 'school days' in virtual classes, the Internet will allow students to spend more time connected to the 'real world'—to professionals on the front lines, working on projects that have real-time effects. Students (thankfully) will not be required to learn 'history' from books, but instead as it happens." —**Brian Reich**, Internet strategist for Mindshare Interactive and editor of the political blog campaignwebreview.com

> "In most respects, I see this as a boon to learning, an opportunity for students from diverse backgrounds to share common experiences, and a fertile field for creative teaching. I fear, however, that these technologies will also enable parents who so choose to circumscribe their children's educational and social environments in ways that fail to prepare the children for diverse workplaces and communities."
>
> —Lois C. Ambash, Metaforix Inc.

"I agree with the second part of this statement. I do not agree with the first part concerning self-paced instruction, particularly for U.S. high school and college students. There is only a limited amount of time available to learn key skills that are essential in later life. The number of those skills seems to be constantly increasing even though the time spent in school is relatively constant. I don't believe it's socially responsible to allow high school students to take basket weaving as opposed to classes that will allow them to read and write and do math at a high school senior level. To do that will mostly condemn them to a life of limited opportunities. The whole function of education is to expand an individual's opportunities. Sometimes that means students need to be encouraged to challenge themselves. If a lot of students are given the choice, they would rather play. Who blames them for that? Unfortunately, a lot of students do not see the downstream ramifications of increasing playtime over work time." —**Robert Lunn**, FocalPoint Analytics, senior researcher, 2004 USC Digital Future Project

> "There will be more choice, but education will still be in classrooms. However, the nature of knowledge and authority are changing rapidly."
>
> —David Weinberger, Evident Marketing Inc.

"There are some situations and uses of 'virtual' classes. But, of all these 'predications,' I feel safest in predicting that the general educational setting will look very similar 10 years from now. Tools will change (e.g., less time in traditional library, more available online). But physical facilities, meeting in classroom will remain predominant." —**Benjamin M. Compaine**, a communications policy expert and consultant for the MIT Program on Internet and Telecoms Convergence

"Learners of all ages will have more tools at their disposal and larger networks of people from which to learn—often without time or place limitations. Lucky ones will even be in communities or professions in which the traditional expectations for judging quality will be liberated. Unfortunately for the rest of us, a short 10 years—even with the rapid growth of more market-responsive for-profit enterprises—will not be long enough to really take advantage of the new forms of learning enabled by the Internet." —**Christine Geith**, Michigan State University

"I disagree only with the word 'most,' as in 'most students.' Some students will do this—students in affluent circumstances and enlightened schools. But 'most' students are not so lucky." —**Gary Chapman**, LBJ School of Public Affairs, University of Texas at Austin

"Student and parental choice is enabled by increasing reliance and adoption of the Internet. Without a doubt, formal education will become more 'customer-friendly' and responsive to student expectations, beliefs, and desires. I do not foresee a future where every student takes an online class—this is too linear an assumption about how the Internet will affect education. Rather, I see every face-to-face class supplemented with collaborative online tools and resources. This blended model to delivering education will challenge the prevailing views of distance education today. There will always be virtual courses, and they will grow in popularity, but they will never be a mainstream part of most students' K–12 education." —**Douglas Levin**, policy analyst, Cable in the Classroom

"This will be increasingly true as the age of the student increases. There will be relatively few virtual classes at the primary level, and far more at the university and adult education levels." —**Jonathan Band**, an attorney specializing in e-commerce and intellectual property with Morrison and Foerster LLP in Washington

"I would like to believe this vision, and it could happen, but pedagogically it seems unlikely. The Internet represents a completely different style of learning. School children and college students would have to learn to be independent, not dependent learners. This requires a huge cultural change. Everything suggests in the U.K. that the government would like to rely on virtual teaching because it might appear cheaper. In reality it will cost more in staff and student time. I suspect that the Internet will be a very helpful resource for education, which might represent a sea-change in learning for a very limited number of students, especially mature students who don't want to attend a campus." —**Nigel Jackson**, Bournemouth University, U.K.

"The technology is there to achieve this; the resources and public will is not. The obsession with standardized test scores is in direct contradiction to allowing students to make their own choices, and it's political suicide to suggest we abandon this obsession. Lack of money and resources will also make it impossible for many schools to take advantage of the technology." —**Rose Vines**, freelance technology journalist, *Australian PC User, Sydney Morning Herald*

"Kids will always be the most creative users of technology. The current classroom setup is just another by-product of the assembly line culture of the industrial revolution, with its neat rows of desks facing the classroom leader (the teacher)."
—**Jonathan Peizer**, CTO, Open Society Institute

"As much as I endorse collaborative learning and student-to-student interaction, I know that many of my colleagues see that as a case of the blind leading the blind. For many, learning is really about the absorption of content, not the making of meaning. For that to change, we need a change in the culture of teaching and learning, not just the technological means. And that will happen slowly, not quickly. Perhaps that's not entirely a bad thing. The student-as-consumer analogy is flawed: Students often learn, not because they want to, but because they are made to. If learning becomes choice-driven, what's to prevent many from making the choice to do less, learn less, tune out?" —**George Otte**, technology expert

> "The educational and digital divides are on parallel tracks. We are headed for an argument between the cost of an education where the student is physically present on campus and the cost of a virtual education...For the virtual class to exist, there has to be a price point that makes it worthwhile to give up the experience of being there in person. If that does not happen, the virtual classroom will be a tactic in a learning portfolio; it will not be the centerpiece."
>
> —Bradford C. Brown,
> National Center for
> Technology and Law

"I agree, although I don't necessarily think that this will help with the development of many skills young people need to succeed. This type of learning is best applied later in the learning process, once fundamental skills are well established."
—**Michelle Manafy**, editor, Information Today Inc., *EContent* magazine, *Intranets* newsletter

"We are pushing hard for the integration of computer and video games into the classroom and think we are making some headway. But schools are among those institutions in our society that are most resistant to change, and your prediction...assumes more radical changes than they are apt to accept. Much of the online learning will continue to be part of the expansion of the role of informal out-of-school learning in student's lives." —**Henry Jenkins**, MIT Comparative Media Studies, author, *Convergence Culture*

> "Learning networks are already becoming a global business, driven by the needs of the developing world. However, the pace of change should not be exaggerated. Many of those attending virtual classes will be sitting in groups in local learning centers, for social and technical support, not in isolation. It will be mixed-mode learning."
>
> —Philip Virgo,
> secretary general, EURIM

"Humans need to interact with other humans in person to learn. Virtual learning is but one piece of the puzzle." —**Joshua Fouts**, executive director, USC Center on Public Diplomacy

"By 2014, an education will certainly include some virtual courses, along with 'classical' courses. In addition, students will be able to go abroad or engage in internships while remaining full-time students. So, the variety of educational experience will be enhanced by use of the Internet." —**Stanley Chodorow**, University of California–San Diego, Council on Library and Information Resources

"This will be only one of the changes education will suffer in the next decade. However, such changes will be more visible and widely implemented at the highest levels of education. The younger are the students, the less people accept changes in education policies." —**Carlos Andrés Peña**, scientific technical leader, Novartis Pharma

"Look at outside factors that could slow this down, such as opposition by teachers' unions fearing loss of jobs. Counter that with the value of accelerated learning (especially where there are not enough teachers or tools), access to remote skills and sources. I'm not sure if the infrastructure (especially the last mile or last few meters) will be sufficiently deployed by 2014." —**Gary Arlen**, Arlen Communications Inc.

"There will still be an important role for in-class education and in-person activities in school. However, these will be enhanced by the use of IT, and, outside of school, young people will continue to develop new and different associations around the globe through sharing of specific interests, skills, etc." —**Ezra Miller**, Ibex Consulting, Ottawa, Canada

"The methods of formal education always lag far behind the possibilities offered by technological advances. Currently, technology promises the chance to tailor education curriculum to individual needs of students. In 10 years, we will see demonstrations of individualized education applied on a small scale, but it will be several more decades before pedagogy makes full use of technology to maximize the individual potential of each student." —**Scott Moore**, Charles and Helen Schwab Foundation

"But this does not guarantee excellence and may lead to a fair degree of shallowness of education based on personal interests." —**Michael Dahan**, Ben Gurion University of the Negev, Department of Comparative Media, Israel

"Do I want to be Plato and bemoan the loss of the great Greek memory skills with the advent of literacy? Maybe. I've been at the forefront of adopting these technologies in the classroom, of trying to implement these ideals. Now I'm stepping back and reconsidering. I'm finding a loss of higher levels of learning (and disciplined learning) in these free-form online versions of the 'open classroom' experiment in physical classrooms in the 1970s. I will always remain a champion of active learning, of overthrowing the authoritarian classroom. But many, when left to their own devices, ignoring even the guides on the side, sandbagging in collaborative groups, gaming the system, are lapsing into higher and higher levels of ignorance, to the point where they have lost the critical thinking abilities to penetrate the logical fallacies and leaps of politicians, to where they fall prey to fascist manipulators of public opinion and become part of an ignorant mob. This is dangerous for sustaining a free society. How can it be that by

> "The educational system will be among the most transformed by nearly free availability of educational material and increasingly sophisticated learning environments."
>
> —William Stewart, LivingInternet.com

challenging the authoritarian nature of traditional classrooms, we leave our students more vulnerable to authoritarian demagogues in other venues? Is this the classic case of the Boomer Hippie parents raising kids who rebel by becoming authoritarian goose-steppers? Perhaps too much loosely structured learning creates a reversal...like McLuhan's media reversals."
—**Christine Boese**, cyberculture researcher, CNN Headline News

"Involvement, yes, but if by civic one means 'proximate or geographic communities' I do not think so. The Internet acts as an intensifier but does not per se create linkages or communities." —**Paul M. A. Baker**, Georgia Centers for Advanced Telecommunications Technology

"In the U.S. at least, this simply can't be the case given the timeframe listed. Reform to the educational system is like turning a battleship. As is, schools are facing economic disparity, a massive shortage in funding for teachers, let alone infrastructure/technology improvements. Add to that the idea of changing an entire curriculum delivery system; it just can't be done in that amount of time. Furthermore, the education of the U.S. youth is still falling behind that of the rest of the world. Virtual classes would serve nothing more than to further the advances of a few while likely allowing the larger group to remain stagnant. While parents and educators stress the need of more individual attention per student, this 'virtual class room' would seem to fly directly in the face of that."
—**Cory Mettee**, Computer Team, Inc.

"Yes, and in the next decade, that's probably more good than bad. The field of elementary and secondary education has really been unable or unwilling to make wholesale changes to its instruction methodology, so it's good to start incorporating new ideas and technology. While there are many who would blame the teachers' unions for slowing the pace of progress here, we can see how the rapid and thoughtless embrace of technology by others has made them unable to now moderate its effects. We'll look back on the upcoming decade and say that the educators got it right, by painfully scrutinizing every incursion into their classrooms." —**Peter Eckart**, director of management information systems, Hull House Association

"As more homes obtain high-speed connection capabilities and lower income households obtain Internet-capable computers, a dramatic paradigm shift from the classroom towards allowing masses of students to tap into top quality learning experiences online will give a whole new meaning to 'home-school.' New technologies will be further refined allowing automated grading, homework will be tailored to appropriately challenge young learners based on where they are as learners and what they are capable of achieving rather than on the lowest common denominator in an overcrowded classroom...Just as technical advances in the manufacturing sector have drastically changed the role of the factory workers and business owners, so too will Internet technology effect the educational infrastructure of teachers, administrators, suppliers, and governmental public education bureaucracies." —**Richard W. DeVries Jr.**, DeVries Strategic Services, St. Charles, Illinois

> "K–12 education is an amazingly conservative institution in the U.S. (I'm not commenting on other countries here.) Local control, even in an era of increased emphasis on federal standards, means that change in school systems is incremental and irregular when viewed from a national perspective."
>
> —Laura Breeden,
> Education Development Center

ANONYMOUS COMMENTS

The following contributions to this discussion of the Internet and education are from predictors who chose to remain anonymous.

> "Education is increasingly moving online, particularly at the university and postgraduate levels. In the next decade, we will also see more effective use of online technology in K–12."

> "Well, maybe. I heard similar predictions for about every communication technology, e.g., cable TV or the French Minitel. There is some evidence that a growing number of students are taking online classes, but I can't tell you that this will be the case for 'most students.'"

"The possibility will provide greater opportunity for a sizeable minority, but active pursuit of knowledge by the majority, I think not."

"Collaboration software is progressing rapidly and classrooms are an ideal application for this technology."

"They already do...much information is gleaned from Internet sources well outside the classroom."

"I hope not. This will have adverse effects on education, which depends heavily upon face-to-face interaction between students and teachers (and among students)."

"Based on my experiences with teaching in virtual teams, students will not take to 'the mastery of their own education.' Most students today, in fact, don't much value learning, but only the degree that they can put on their resume. Left to their own devices, most students would do significantly less academic work."

"The evidence of learning and of learnedness is to the contrary. There is nothing about the medium that will lead this way and more evidence that IT-in-education is a colossal failure."

"Technology can only serve ends set by those who control it—I see no reason to believe that educators will choose to use the technology in this way, though it could have powerful results if they did choose to do so."

> "It didn't work in the past with other technologies, it hasn't worked with the Internet to date; in fact, distant education was always considered a second best to being there—current research supports these findings again with the Internet. ICTs will be a resource, not a replacement."
>
> —Anonymous respondent

"This will occur to some extent. Experience indicates that the time involved in producing quality virtual learning materials is high. Furthermore, one hopes that the best scholars are producing the learning materials as opposed to people with IT prowess."

"Let us hope this is our future. It cannot happen unless there is the political will to make it happen. So far, the evidence from the U.S. DOE [Department of Education] and various state DOEs is dismal."

"Not likely. Pace of change and innovation is slow. We are already in the 'third decade' of the so-called 'computer revolution' in education, yet there has been too little change in the majority of American classrooms. What will change, however, is the role of the Web and Internet as a critical supplement to the activities in the classroom, both K–12 and college."

> "Lack of money and a lack of commitment to this sort of goal—and the inequity of funding in education, at least in the U.S.—makes this unlikely."
>
> —Anonymous respondent

"Fortunately or unfortunately this will probably be the case. Online universities are already doing this; it is only a matter of time before it shifts to primary and secondary education. Unfortunately, such means of education, while excellent at the purely pedantic level, simply do not foster the development of social learning that exists in a more traditional environment."

"I would like to see the [provided] scenario, but having been a professor for 10 years, mainstream education is very slow getting off the mark. Much of this change will come outside of school."

"This is already beginning to take place, although generally not in the public school system that is bent on getting their school a good rating on the next assessment test."

"Student behavior depends on change overtaking the educational bureaucracy. In other words, our human technology, i.e., how we agree to teach children, has to change before the kind of change you are talking about can come true in the classroom. Students already are learning tremendous amounts virtually outside of the classroom, much of which is perhaps not what teachers would have them learn."

"Increasingly, the technology will be there to permit this, but educational institutions are very slow to change, and the result is that we still will make relatively little use of the capability except in certain niche areas."

"This will be true for a number of students but not most. Education will never move that fast."

"When it comes to changing pedagogy, a decade is nothing. We will go on as we are for at least 50 years before things change significantly for most people. Some early users will do some changing in the next decade—but not many."

> "This is already unfolding. Technologies for online learning are being brought into the formerly offline classes. More formal aspects of learning (and thus also of teaching) will also extend out of the classroom and class time to fill the week. Indeed, this is already here."
>
> —Anonymous respondent

"The traditional role of teacher will be diminished as students find peers and other authority figures from which learning can be done. The downside of this is that traditional training for teachers will be inadequate to deal with this kind of classroom, or how to exploit this kind of learning/teaching dynamic. Only a small portion of 'school time' will be devoted to this learning environment, as it has traditionally been viewed as 'extracurricular.' Little value can be placed on the knowledge and skills attained with this learning because there exist no standardized measures of what students learn and apply outside the traditional classroom. There is potential for growth within the area of formal education, but little initiative or consensus on how to implement it."

"There are limits on the efficacy of student-directed learning, and I think they have already been discovered. The second statement seems more plausible."

"Already happening. One of our local education authorities has deconstructed the curriculum in some of its schools and school kids pace themselves, using notebook PCs. The educational experience will be much more diverse and there will be a general up-skilling. For disadvantaged groups, including those discriminated against, such as young black boys, the impact will be very positive."

"The educational model will evolve to lifelong learning over degree institutions."

"Rather than 'freeing' students, the technology will be used to make mass education cheaper. For a small proportion, technology is a dynamic extension of an enriched educational process. But for many, it is simply a more efficient and economical way to deal with a burdensomely large student population with far too many needs."

"I'm not sure, and this is pretty much where I spend most of my professional time. The instructional apparatus is very tenacious in protecting itself. And the K–12 teacher scaling issue is deadlocked (i.e., we can't reduce the student-teacher ratio much). I think it'll take another decade for the instructional population to turn over."

> "Students will continue to require face-to-face mentoring. Part of school is social development. Graduate work will be almost entirely virtual."
>
> —Anonymous respondent

"Learning is very difficult to do online. There would have to be very rapid change and development of the educational sphere for this to happen in the next 10 years. A lot of schools still don't integrate information technology into their physical infrastructure or curriculum."

"They might want to do that, but our crumbling educational system is a sluggish, bureaucratic morass that cannot figure out how to budget for anything let alone innovate."

"The millennium generation will interact and socialise in radically different ways—this will have a major impact on the experience of education."

"The need for formally sanctioned learning means traditional methods—and the power of professionals rather than students—will dominate. However, students will spend more of their study time outside of class in such virtual fora and very little in libraries or reading printed materials."

"I am an educator. I regularly develop and teach online and partly online courses, and am fully convinced of their relevance and utility in some situations. But I do not believe that they will displace regular classes and real-time face-to-face interaction with teachers as a preferred mode of learning, especially in some content areas."

"The trend toward directing learning according to student choices long predates the modern IT. What the Internet changes is opportunities for students to learn from people at a distance. The 'teachers' in question may be other students, but may also include role models, mentors, and teachers elsewhere."

"While I have no doubt that the technology permits and potentially could optimize virtual education, I do not see the conservative core of the country (U.S.) approving use of tax dollars for this purpose. I now live in a state that has an appalling policy that values a zero-based budget over necessary public education expenditures in the present let alone investments for the future."

"I find it extremely difficult to believe that our educational structure in the U.S. could do something as revolutionary as allowing students to grow at their own pace. It would take a complete revolution in our educational system to make this happen."

> "Even if we limit ourselves to the U.S., we're simply too far away from a society where 'most students' have access to computers even some of the time. Much less where entire curricula could be designed around networked learning for anything more than a small fraction of the student population."
>
> —Anonymous respondent

"By 2014, many of the wealthier suburban and private schools may well be in this 'advanced' mode, but I don't think it will yet be the norm for 'most students.' But the prediction itself isn't aligned well. 'Set by student choices' could mean a haphazard pursuit of immature interests and fads or a carefully structured sequence of authentic learning experiences and inquiry learning projects that may or may not require online activity (e.g., the Anderson School of the Future in California is highly innovative in this regard but used little technology). I happen to believe that such a restructuring should occur and will be best enabled by strong doses of virtual/online interaction, but it will take longer than 10 more years to get even the majority of schools headed there. And we cannot afford to continue to widen the achievement gap for lower-income and racial/ethnic demographic groups of young people."

"Students may think at first this is a good idea, and they may enroll in these virtual classes, and they will likely try one out, but the fact is (and I have taught and talked to a lot of learners in this environment) the only ones who can 'learn' in this environment are: (a) those who can read very well, (b) write articulately, and (c) are highly motivated to learn. That is limiting, to say the least. Plus, most people want personal contact and group interaction in a learning environment. Until online audio-video real-time contact is generally available, virtual classes will not appeal to the majority of learners. It will be fine for 'mature' learners who are highly motivated. For the average, it will not work. The vast majority of youthful students I speak to tell me that they would far prefer a 'regular' class to a virtual class. For those who are at the MBA level or who are mature students who are holding down full-time jobs, I agree that they would prefer the virtual classroom, simply because it is the only realistic choice they have based on their situations, but even they would prefer a more traditional classroom. Maybe when the technology is ubiquitous and everyone can have broadband and all classes can be simulcast it may work. In the meantime...virtual classes will remain the last option for most students."

> "People want to learn in social settings. The advancement we finally are seeing today is that the Internet is facilitating social interaction. It's all becoming easy to use and rich in terms of what can be conveyed and shared. I fully expect this to continue on a reasonable, gentle growth curve until computer-mediated interactions are a significant natural part of a learning experience—alongside the traditional classroom setting."
>
> —Anonymous respondent

"These are two very different items within the one question—I certainly don't predict education being increasingly driven by student choice, but increasingly by federal government directives that focus on demonstrable, relatively low-level skills. But, I also see more and more students working online as part of their coursework—whether as a homeschool homework interface, content delivery during a school day, or substitution of some classes with online discussion sessions—but much of this will be driven by economics (i.e., fewer teachers can teach more students using online media) than by, say, research showing students learn more effectively online. Education—especially compulsory education spheres—has a woeful track record in taking up new technologies in meaningful and useful ways."

"Schools are slow to evolve, and in-person student/teacher learning will remain the norm. Virtual classes are useful, but I don't believe they'll become the norm in the next 10 years."

"Students will never control their in-school choices and this is the largest threat to national competitiveness and security. We have taken away the teacher's sandbox, his arrows, and quiver. We are failing at education. Children have self-organized outside of institutionalized education to produce the KSA of future workforce needs. Network video-game builders are engaged in transdisciplinary, inquiry-driven, self-motivated learning. They are creating new worlds, new processes, new techniques, new languages, and new knowledge. We can not seem to pierce the veil of their play to understand their learning much less their attitudes, beliefs, and aspirations. Generation Y is the architect of global futures. Can we trust them? Can we entrust them? Can we trust ourselves enough to let them go? To be free? To explore? To invent? We are experiencing a renaissance. Institutionalized education is lost to industrial and agrarian structures, influences, and perceptions. We need a qualitative transformation of learning from the students up. Teachers are students and those who get it are in it."

Part 9

Democratic Processes

Prediction: *By 2014, network security concerns will be solved and more than half of American votes will be cast online, resulting in increased voter turnout.*

Experts' Reactions	
Agree	32%
Disagree	35%
Challenge	15%
Did not respond	18%

Note. Since results are based on a nonrandom sample, a margin of error cannot be computed.

It is safe to interpret these results as a clear challenge to the hope that online voting will become normalized anytime soon. The experts who disagreed or challenged this prediction articulated diverse concerns, often citing their professional expertise. Few of those who agreed with the prediction shared justifications beyond hopeful bromides like, "The sooner the better."

Many experts sought to crush the optimistic view that network security concerns could ever be "solved."

Many experts were frustrated that the prediction mixed too many elements. One expert wrote, "Many parts to this prediction. Security concerns are unlikely to be solved and so people will not want to vote online. Even if they could, though, I do not see this resulting in increased turnout. Also, what about the digital divide? It is not going away in the next 10 years. That will suppress online turnout in many sectors of society."

The ritual of the voting booth attracted some experts. Douglas Rushkoff, an author and professor at the New York University Interactive Telecommunications Program, wrote, "I think people will begin to devalue voting if they don't go do it somewhere. So, increased access may lead to decreased participation."

J. Scott Marcus, a senior adviser for Internet technology at the Federal Communications Commission, wrote, "'Solved' may be overoptimistic, but at least ameliorated to the point where deployment is realistic. Increase turnout, yes. And also possibly facilitating more advanced forms of voting, if states/municipalities are willing to try them. (But the U.S. has been resistant to systems such as proportional representation—it's too democratic.) ;^)"

Philip Virgo, secretary general of EURIM, wrote, "And, if so, America will have ceased to be a democracy. The problem is not merely 'network security' or even equipment security, but that without a secret ballot, monitored by representatives of the candidates and/or independent observers, there is no reliable way of preventing coercion (even if there were ways of preventing impersonation). Voting is a social, not a mechanistic, activity."

> "As long as one human being designs it, another will be able to break or hack it. This is one truism we can never lose sight of."
>
> —Anonymous respondent

One expert looked beyond voting: "I am contaminated by the current norms of two-party political behavior. I see nothing but continued estrangement from national/conventional politics. But, at the same time, I believe virtual communities of interest will exercise episodic political power...like a swarm of angry bees!"

ADDITIONAL RESPONSES

Many other survey respondents shared valuable remarks in their elaborations following the question about democratic processes online. Among them:

"There's a good chance that in 10 years we will have learned to design robust, trustworthy voting systems. But voter apathy is related not to the voting system but to the perception that the vote counts. Politicians of both parties have so manipulated the system that only 29 of 435 voting districts have any sort of contest this year. It's very easy for many voters to become apathetic when they perceive that their vote will not affect the outcome." —**Peter Denning**, Naval Postgraduate School, Monterey, California, columnist, *Communications of the ACM*

> "Here's my revision: By 2014, network security concerns will still be with us, more than half of American votes will be cast online anyway, and this will have no effect on the rates of voter participation."
>
> —Anonymous respondent

"Network security concerns will never be solved, though the trend toward increasing use of online voting will continue. The very cost-effectiveness of this technology encourages governments to adopt it." —**Jorge Reina Schement**, director of the Institute for Information Policy at Penn State University

"The great voter riots of 2008 based on distrust of e-voting machines will stifle online voting." —**David Weinberger**, Evident Marketing Inc.

"First, network security concerns can be sufficiently resolved without intergovernmental cooperation of a kind that is unlikely. Second, the ritual of casting one's ballot plays an important role in voter participation. I fear that if we can vote at our home computers, voter turnout might increase only among the most committed and best-organized interest groups. Third, many politically active people, myself included, will oppose online voting and computer voting of all kinds until we are assured of a viable and legitimate paper trail. I think it would take more than 10 years to put the political and legal mechanisms for online voting in place, even if the technological issues were resolved tomorrow." —**Lois C. Ambash**, Metforix Inc.

> "It is foolhardy to underestimate the fragility and vulnerability of any online system to attack and manipulation. Anything that can be made secure can be hacked."
>
> —**Mack Reed**,
> Digital Government
> Research Center, USC

"Voting security is likely unobtainable, regardless of the technology. There is too much at stake, and there are too many incentives to corrupt the process. There will need to be a physical representation of a vote in the future." —**Ted Eytan**, MD, Group Health Cooperative

"Electronic voting will not raise turnout because the difficulty of casting a ballot is not a reason for low turnout. Furthermore, the basic anonymity of a ballot is very difficult to protect if people may vote from home. I don't see a technical fix for this problem." —**Peter Levine**, deputy director, Center for Information and Research on Civic Learning and Engagement, University of Maryland

"Even with the security concerns being solved, it is difficult to predict a turnout increasing since the current low rate is not directly related to the electoral tools. There are other reasons affecting the whole political system that may better explain this situation." —**Jordi Barrat i Esteve**, Electronic Voting Observatory, Universitat Rovira I Virgili

"By then there will be a serious security layer added to the network model because politics and economics will demand it. To what extent this also facilitates surveillance and censorship is still unclear. If Americans can vote online, turnout will be increased—but the extent to which this occurs will depend on voter scandals traced back to digital technology facilitating voter fraud." —**Jonathan Peizer**, CTO, Open Society Institute

"I don't think this will happen quickly. Too many people get too much joy from suppressing voters at physical polling places. It would take a political tsunami to make this happen."
—**Susan Crawford**, a policy analyst for the Center for Democracy & Technology and a Fellow with the Yale Law School Information Society Project

"Votes will be cast online, but if current voting technology is any indicator, we'll have no assurance of the security of the technology. If that's the case, concerns about the integrity of the system may depress its impact on voter turnout. And unless there is a major public policy initiative that places a networked computer in every American home (a very unlikely scenario, I fear), online voting will make voting easier for the social groups who are already more inclined to vote and leave behind those who are already disenfranchised." —**Alexandra Samuel**, Harvard University, Cairns Project (New York Law School)

"I challenge this prediction. You not only have network security concerns, you also have personal identity concerns (different but highly important issue). You solve a lot of identity problems by making a person come in to a centralized facility to vote. Getting that process computerized is the next logical step. Only after that can we consider the next step. I seriously doubt 50% of Americans are going to be electronically voting within the next 10 years. I don't know about the increase in voter turnout, either. I question the implicit assumption in this prediction that making voting electronic will automatically increase the number of people voting. I suspect one of the primary reasons for low voter turnout is a perceived lack of personally relevant choices that the people have. This is very likely due to the two-party political system in the United States. What I see happening is that a lot of campaigning is going to be online. Once that happens, the two-party political system is going to dissolve. That

> "Online voting will become at least as secure and reliable as traditional voting. Although, I'm sure we can strive for a higher standard."
>
> —Brian Reich, Internet strategist for Mindshare Interactive and editor of the political blog campaignwebreview.com

should lead to more choices, and that should increase voter turnout. In other words, what I see happening is that the entire political power structure of the United States is very likely going to change as a result of the Internet. However, that is not going to happen in 10 years, either." —**Robert Lunn**, Focal-Point Analytics, senior researcher, 2004 USC Digital Future Project

"This prediction could only come true if there was an American standard for voting. It isn't only network security concerns that will drive this opportunity. It is a long list of special interests, local political situations, and lack of voter education. What happens outside the network security is vastly more complex and important. For example, when the software vendor sends an update CD to a local voter administrator, who's to say that the administrator won't take the CD home and melt it in the microwave for a planter? Certainly, a voter has no idea whether or not the software that supports the vote transaction is up to date, built with integrity, or backed up to the extent that a vote cast will be a vote counted." —**Elle Tracy**, The Results Group

> "No, there will always be new risks—technology has never been and will never be foolproof and error-free. Increased voter turnout is not guaranteed as: (a) many people continue not to vote in spite of the increased options available to them—once again it's a personality and not a technology issue (technology only gives more freedom if an individual is predisposed to accepting more freedom); (b) technology problems could still lead to inaccuracies in vote counting!"
>
> —Bornali Halder,
> World Development Movement

"It is much more likely that mail-in ballots will become the norm. The voting system is very change resistant and power is distributed among many nodes." —**Ted M. Coopman**, University of Washington

"The reasons for low voter turnout are many and complex and are not, in the majority of cases, to do with the effort involved in voting. The Internet may make it easier to vote, but it may not solve issues of disaffection, alienation from the political process, feeling that votes don't count, that voting doesn't

give you a voice, being ill-informed about politics, feeling that there is no real choice within the current constraints of (in the U.S.) the two-party system. The Internet is a tool—you need to understand why other democratic tools 'aren't working' in order to understand how this tool might affect voter behavior."
—**Susan Kenyon**, University of the West of England, U.K.

"Beginning with (the 2004) election, challenges to the accuracy and safety of electronic voting will become a major concern, which will take several decades to resolve." —**Peter W. Van Ness**, principal, Van Ness Group

"While network security concerns may well be solved, I don't expect to see widespread trust in the fact that they have been solved. That is, to really 'work,' an election must be perceived by the populace as having been fairly conducted and fairly counted." —**Thomas Erickson**, IBM Research

"Add open-source software, a public agency that manages the systems, the requirement to attend a polling station, and a verifiable paper trail, and most electronic voting systems can be made to work, and probably will be. But voting online from home, without having to show up in bodily form at a local location, is a system that cannot be secured." —**William Stewart**, LivingInternet.com

> "**BWAH-HA HA HA HA HA HA HA!!!** 'What? Would you then disenfranchise the noble dead?' —A statement by 'Congersman Frog,' in the cartoon strip 'Pogo,' by Walt Kelly."
>
> —Mike O'Brien, The Aerospace Corporation

ANONYMOUS COMMENTS

The following excellent contributions to this discussion of democratic processes online are from predictors who chose to remain anonymous.

> "This will no doubt come, but it has already surprised me how slow the evolution in this area has been. I don't think we will get to 'more than half' in 10 years. Maybe 20."

"Network security is the biggest unsolved issue for the Internet. There is no indication that security issues will be solved. The inherent conflict between privacy and security remains unsolved."

"No doubt about it. Make it easier, vote from home, send an e-mail reminder and more people will vote. This will happen faster than 2014. It will make the campaign process more interesting and more interactive. We can all be virtual delegates."

"I do not believe network security concerns will be solved—ever. There will always be threats. As old threats are mitigated, new threats will emerge."

"Online voting in Europe has not been shown to result in increased voter turnout."

"The democratic political process should not be left to the Internet, and I do not believe that elected and appointed officials responsible for running elections will take the risk in the next decade… it could make the 2000 presidential election look like a cakewalk."

"Not all security concerns will be solved. I believe that there are nontechnical objections to online voting that are fundamental, for example, the facilitation of vote-buying. Other sorts of voting (such as share-holder meetings) will be totally online."

> "So far, rate in changes in voter technologies has been slow, public distrust is high, even in the face of major scandals after the 2000 elections. I am not sure whether moving voting online is a good thing in any case given the inequalities of Internet access. It could result in the disenfranchisement of a significant segment of the population if the digital divide does not get resolved."
>
> —Anonymous respondent

"So far, rate in changes in voter technologies has been slow, public distrust is high, even in the face of major scandals after the 2000 elections. I am not sure whether moving voting online is a good thing in any case given the inequalities of Internet access. It could result in the disenfranchisement of a significant segment of the population if the digital divide does not get resolved."

"Although voting may move online increasingly, security issues will not have been solved (current ones maybe, but new ones will keep arising), leaving us with a very vulnerable and corruptible system."

"Security concerns will be ignored. Votes will be cast online, but public-choice theory suggests there is no reason that this should increase voter 'turnout.'"

"Whether or not the security concerns are solved, voting will move online. Pressure from citizens, as well as key legislators, will make that happen. It could do one of two things: Help conquer the digital divide or make it grow ever wider. It will all depend on access."

"Politicians will argue against anything that will get greater involvement."

"The prediction implicitly assumes that online voting is held up by network security concerns. While I believe that network security will vastly improve in the coming years, I don't believe that voting security will improve. Online voting is open to very much the same fraud patterns as absentee ballots, with one aggravating circumstance: It can be automated. Network security will not make absentee ballots safer."

"Considering the travesty of the last election and continued reports about campaigns of disenfranchisement for this upcoming one, I think we will have to go a long way before people give unquestioning trust to an all-digital system. Furthermore, both Democrats and Republicans actively benefit from low voter turnout in that they have smaller target groups to convince each election year. The fact that we are only watching presidential commercials in a handful of swing-states this election surely says something about the parties' desire to have the entire electorate vote. That said, the convenience of the Web speeds progress and diminishes cost."

"If we want to, this is a solvable problem, and by 2014, we could do it. I don't know if we have the will or the resources, though."

"Trust mechanisms are just starting to be explored. The infrastructure necessary to support this kind of widespread authentication and authorization policy and operation will only just be making real progress at this level."

"Security concerns of 'Internet voting' cannot be solved. Those who claim otherwise lack very basic understanding of the problems associated with voting."

"Network security will always be an issue, as there are always 'smart' people out there trying to sabotage the network. Votes? The same people who won't vote, won't, and the 30% of the population without online capabilities will not be able to vote unless there are polling stations."

"I'd say that substantial number of votes will be cast online, although not 50%. Security problems may be contained but certainly not solved!"

"I don't see information assurance as having a 'solution.' It is an evolving struggle between attackers and defenders. I don't see anything changing that in 10 years."

"There's always a better mousetrap that can be compromised by determined individuals bent on disrupting the Internet. Not everyone is computer literate and will vote correctly. There will be some sort of hanging chad in cyberspace."

> "Solving the problems of online voting will prove very difficult, and there will be resistance at almost every level to the implementation of these systems."
>
> —Anonymous respondent

"If more Americans vote, it will be because one of the major parties manages to distribute an application that does your voting for you, as directed by the party; disengagement with the world is not solved by tech toys. Issues are complicated, and voter pamphlets require time to digest. That's the barrier to entry."

"We have some hard thinking here about how to have elections that are not rippled with fraud. We will have to consider whether the anonymous vote continues to be viable. But if it is not, what then—how can we have accountable elections but anonymous votes? Something may have to give."

"Voting will never be done via Internet for a large portion of the public."

"The people who benefit from the current electoral system will not allow it to change. 'Ballot security' concerns will trump technological advances."

"There will always be hackers and workarounds. If network security concerns are ever solved in such a way as to allow absolutely no possibility of voter fraud, I doubt it would happen within the next 10 years."

"Even if it could be done, I don't think a majority of the population would trust it."

"(1) Network security concerns will never be solved. The hackers will always be one step ahead of the good guys. This is not an Internet issue, it's a human/organizational nature issue. (2) There are so many more obstacles to online voting than network security, that they cannot be resolved by 2014. (3) Every report of a failure, however small and however quickly fixed, will result in decreased trust in the system; hardly conducive to greater turnout. There are many other actions that would increase turnout more predictably and with fewer drawbacks than Internet voting—a uniform 24-hour voting period, for example. On the other hand, we have good evidence via research and e-commerce that if people really want what the Internet offers, they will hold their noses and plunge ahead. If Internet voting can be made good enough, it could have a positive effect on voter turnout, but not a dramatic one."

> "This is a no-end battle. A good security engineer is sure to have work all his life. The question being not how to make a secure system but: How long will it live before being hacked? This has no impact on voting."
>
> —Anonymous respondent

"Network security concerns won't be 'solved'…ever. Folks might vote online but this will not necessarily increase voter turnout."

"There will still be too much social suspicion of technologies (and more importantly, our political process) to see e-voting by 2014…perhaps further down the road, but not by 2014."

"I agree with the first part, but not the second. Sooner than 2014 it will be safe to cast votes online. However, it will not increase voter turnout because there will be a single party after the election of George W. Bush to a second term. Redistricting will continue at a fervent pace beneficial to Republicans, the Democratic Party will collapse, and America will 'evolve' to a one-party plutocracy/Christian theocracy that represents the values of approximately 25% of the American public. By 2020, America will begin to resemble early 1990s South Africa. Wealth and power will be concentrated among corporations and a small number of individuals (even more so than it is today) and the majority of the American people will be disenfranchised."

"We're going to an online vote for fast results, verifiable by a paper ballot, where elections are close. I don't think security is ever going to be that good. Some people don't trust government. Why do you think there are guns?"

"I would point to IPv6 as the solution and key enabler."

"Not only must network security be solved, which I do not believe you will ever have complete network security, you must also solve the issue of online authentication."

"[I agree with] all but the network security, which [is a problem that] will always be with us. We live in a world of 10 levels of device iteration. Not all can own the latest technology all at once."

"There is very little proof that ease of access beyond a certain point actually increases voter involvement. Removing actual barriers to voting certainly has that effect. But in the U.S., where all you have to do is remember where and when to vote and then show up, there will not be a dramatic increase in voting."

"Let's hope so. But networks may be the least of the security concerns. Take that leaked Diebold letter that promised to deliver Ohio to the GOP. This would be a great leap forward, so long as there isn't a finger on the button."

Part 10

Families

Prediction: *By 2014, as telecommuting and homeschooling expand, the boundaries between work and leisure will diminish significantly. This will sharply alter everyday family dynamics.*

Experts' Reactions	
Agree	56%
Disagree	17%
Challenge	9%
Did not respond	18%

Note. Since results are based on a nonrandom sample, a margin of error cannot be computed.

Home broadband users and teenagers may represent leading indicators for what the future holds for the intersections of work, school, and home life. For example, Internet users who have added a fast connection at home say they are more likely to telecommute. A home broadband connection also appears to make it easier for family members to share access to the computer and the Internet.[1] High school students see the Internet as a virtual textbook and reference library; as a virtual tutor and study shortcut; as a place to conduct virtual study groups; as a virtual locker, backpack, and notebook; and as a virtual guidance counselor when they are deciding about careers and colleges.[2]

Since most of the respondents to this survey are elite "knowledge workers"[3] with fast connections to the Internet at home and at work, many shared personal stories about what a difference the Internet has made in their own working lives.

One person wrote, "The increase in connectivity between mobile devices will result in a new family dynamic that will re-expand the notion of family to include not only geographically displaced extended family relatives but also unrelated family members. Around the clock connection and automatic sharing of contact information beyond the immediate family members will foster digital tribes and a stronger sense of 'family.'"

Another expert wrote, "This prediction assumes that the boundary between work and leisure is a natural thing. On the contrary, it's a new way of organizing. Moving back towards more integration will happen and is a good thing."

Many respondents who offered up their personal experiences as proof were outshined by skeptics who, while in the minority, wrote convincing arguments to the contrary. One expert wrote, "I think it is naive to view the technology in isolation; family dynamics [as well as] economic and social considerations, will trump this prediction—10 years is too fast to expect such change. Perhaps in 2 decades, but this extrapolation does not seem consistent with America's past tech influences on lives."

> "The Web is dramatically changing the way women in my generation are able to mother and work. The Web is providing the tool that women needed to contribute at home and in the world."
>
> —Tiffany Shlain,
> founder, the Webby Awards

Another person wrote, "The potential exists, but the time saved by not commuting isn't likely to be used to increase the workday. There is no evidence that kids living in dorms or boarding schools can't separate class time from playtime—even if they're living within 100 yards of the classroom."

Two respondents expressed doubt that all workers would benefit from telework options. One wrote, "I also remember when videoconferencing was supposed to end the need for business travel. Sometimes there's no substitute for being there." Another pointed out that "nobody is going to make cappuccinos in Starbucks from home."

> "Whoever suggested that homeschooling would increase because of the Internet has never stayed at home with a child. The Venn diagram of telework and homeschooling shows two circles entirely without connection."
>
> —Moira K. Gunn, host, *Tech Nation*

ADDITIONAL CREDITED RESPONSES

Many other survey respondents shared incisive, overarching remarks to the question about the Internet and family dynamics. Among them:

> "That's already happened. It's all work. Even shopping."
> —**Douglas Rushkoff**, author; New York University Interactive Telecommunications Program

> "[It] already is [altering everyday family dynamics], as Wellman and NetLab expect to find in their current Connected Lives study." —**Barry Wellman**, University of Toronto

> "I would not be surprised to see a backlash as family dynamics suffer from the 'on, all-the-time, syndrome.'"
> —**Jan Schaffer**, J-Lab: The Institute for Interactive Journalism

> "I would revise and specify this prediction: By 2014, as convenience computing brings the Internet into more moments and arenas of everyday life, the boundaries between work and leisure will diminish significantly. This will increase the power of corporate interests in determining how people frame the public and private arenas. This will also cause an overall increase in levels of stress and fatigue in the general workforce."
> —**A. Markham**, University of the Virgin Islands

"Some of this is already happening: Many workers now are 'on duty' 24/7—responding to e-mails, alerts, BlackBerries, and cell phones, no matter where they may be. For the office, this may increase productivity. For the home and family, this adds to stress and strain. But that is because, today, this 'extra' duty usually comes on top of a regular 40- to 50-hour stint in the office. In the future, it will be possible for people to do their work from home, from the beach, from the backyard—and it will be theoretically possible to enhance home and family that way. Again, it's not the technology that will decide this; it is our institutions and their rules." —**Gary Bachula**, Internet2

"I believe people will learn to understand how their different identities (home, work, others) are represented online, and new tools will be developed to help people maintain healthy boundaries around work and leisure to maximize health."
—**Liz Rykert**, Meta Strategies Inc., Toronto, Canada

"I could not more strongly agree. I think that we are already seeing that the greatest change that the Internet enables is a fluidity of task over space. There is a tradition in many professions (including my own) of flexible workdays and places. I suspect that this flexibility will increasingly affect all forms of knowledge work, and this will be felt most acutely in our social and familial organization."
—**Alex Halavais**, State University of New York at Buffalo

> "I guess this will mean there is a reason for four-person families to be living in all these minimansions. By 2014, they will need the space so they won't kill each other!"
>
> —Clare De Cleene,
> Administrative Office
> of the U.S. Courts

"I see it in my family every day. Sure, it is anecdotal evidence, but it is powerful."
—**David Tewksbury**, University of Illinois at Urbana-Champaign

"The boundaries between work and leisure will continue to diminish, but I don't think they will change much more from where they are today. Most employers will want most employees on site most of the time." —**Jonathan Band**, an attorney specializing in e-commerce and intellectual property with Morrison and Foerster LLP in Washington

"This has already happened. Everything is a hobby—half work, half leisure—it's an unstoppable trend. People forced to be offline feel spiritless and lonely. We're there." —**Susan Crawford**, a policy analyst for the Center for Democracy & Technology and a Fellow with the Yale Law School Information Society Project

"While I think the move toward telework has been slower than anyone expected, the growth of homeschooling has been faster. Every time I speak before the public, most of my questions center on the impact of these technologies on family life. People are concerned about and aware of these potential changes. For the most part, they are very nervous about a world where it is impossible to escape the office and where they face growing competition for their children's attention." —**Henry Jenkins**, MIT Comparative Media Studies, author, *Convergence Culture*

> "What we need to go along with this trend is a new definition of quality work—judged by outcome, not by time. Education needs a similar redefinition."
>
> —Christine Geith,
> Michigan State University

"Many Americans (and I am as guilty of this as anyone) work too much and carry their stress home with them. It will become too easy for Americans to work and play at the same time, likely leading to some diminishing of both." —**Brian Reich**, Internet strategist for Mindshare Interactive and editor of the political blog campaignwebreview.com

"My greatest hope is that telework will hit the federal government in a big way. There's absolutely no need to have those huge headquarters' operations located in Washington, D.C. It's an expensive place to be, it's a limited applicant pool for jobs, and it creates a huge 'sitting duck' for terrorists. Most of those operations could function beautifully scattered throughout the country, using workers in their homes and in telework centers, working virtually. In fact, 3 years ago, I moved from Washington, D.C., to Tucson, taking my job as departmental Web manager at HUD with me. While there are many struggles—mostly overcoming people's reluctance to work online—we have proven that this can work quite successfully." —**Candi Harrison**, Web content manager, U.S. Department of Housing and Urban Development

"Many forces already conspire to alter family dynamics. I don't see this as any more powerful than the divorce rate, single parents, two-income families, wildly fluctuating economy, rising gap between rich and poor, etc." —**Peter W. Van Ness**, principal, Van Ness Group

"This change to family dynamics can be very positive. Just as writing skills will become as important as they were early in American history, we can return to the home-centered work environment. When farms were the center of American life, families were an integral part of the workday. As we remove boundaries between work, personal, and family life, families can grow closer and participate more with all aspects of life. This will impact education as well. Separate school systems and activities all day remove kids from the day-to-day activities and decisions about how mom and dad actually pay for things. Having them more involved will change the topics they are interested in and the perspective they bring to the classroom."
—**Mike Witherspoon**, Connexxia

> "Work will still be work, and leisure still will be not working. The places where they happen—especially work—will shift. It will be possible to move between the two more quickly, but the people who make work will know this and raise the productivity bar. It will still be a matter of personal choice whether one agrees to the work contract offered or not...The Internet causes power to disperse, in this case, to the people from the institutions. Of course, institutions can attract power (the will of the people) but they must do so under new rules of engagement."
>
> —Mike Reilly,
> president,
> Hally Enterprises, Inc.

"Of course, this is already happening for many of us. My workdays begin online at 5:30 a.m.; I am out in the countryside on my bicycle by 8 a.m.; I am in the office by 10 a.m. (both online and in face-to-face meetings); I am out of the office by 5 p.m.; and I am back at work online from home at 7 p.m. I teach my

classes online from distant points. This semester alone, I will have taught classes (engaging students at least twice a day) while off at weeklong conferences in Boston, Chicago, and Orlando."
—**Ray Schroeder**, University of Illinois

"I would agree in that this mirrors our own family life. My wife and I work as consultants from home and we blend work and kid time as both require. However, I remember 40- to-60-hour-a-week jobs that required me/us to be on site as staff. I really don't see that changing much at any time in the future. There are valid reasons telecommuting has not taken off. What might change is the ability to see/reach family during work hours in ways that save time and absenteeism. You might be able to e-mail your dry cleaner to drop your clothes in a box outside your house. Or have a little window open on your computer to watch your kid in daycare. But a lot of this is happening already." —**Tim Slavin**, ReachCustomersOnline.com

"Homeschooling is not going to expand by an order of magnitude, because most parents don't want to do it, and some of those that do, can't. So the kids are going to a physical school, even if they do a bunch of online stuff once the get there. And telework might increase, but in the next decade we're still going to see most workers leaving their homes to go to work. They will get some hours or days to work from home, but their primary workplace will still be an outside location. The boundaries between work and leisure will be more threatened by the fact that people use their work-Internet access to do stuff that's not work. We'll see a major crackdown on how the Internet is used at work, now that the technology to monitor and block nonwork activities is maturing." —**Peter Eckart**, director of management information systems, Hull House Association

> "The same segments that are today high-volume consumers and television- and media-centered have already yielded most of their family dynamics, so this will not likely change for them. Another small segment that today is not highly-penetrated by media will see their family dynamics affected."
>
> —Dan Ness, MetaFacts

Anonymous Comments

The following excellent contributions to this discussion of the Internet and family dynamics are from predictors who chose to remain anonymous.

"This is already happening." —*Anonymous response from dozens of participants*

"We live in a world that is always on. I think this will be one of the most devastating consequences of technology."

"I agree. I just started working from home and I can't get away."

"This will be true for a significant number of people, but only a tiny fraction of the overall population."

"The boundaries between work and leisure have already been erased. Everyday family dynamics have already been changed."

"Computers in the home mean work in the home, and Internet connections mean demand for connection from home. Not a future scenario, but a current one. Family dynamics will be pushed by this, but will it be better or worse than the change that put two parents in to the workforce to make ends meet?"

"Over the next decade, families will need to increase work hours dramatically in order to keep up their standard of living. We will have moved from the one-wage-earner family to the two-wage-earner family to the multiple-wage-earners-with-multiple-jobs family. As a result, families will be increasingly scheduled, to the point where a family member will be working almost every hour of the day."

"People work for social interchange. Kids go to school to learn to get along with their peers."

"Already true today. Technology allows work time to expand to 7 days a week."

"I believe this prediction is already coming to fruition. A survey would likely determine that most Americans check their work e-mail from home. Laptops are packed with golf clubs and snorkels for family vacations; BlackBerries and cell phones are commonplace on sidelines and in stands."

"This is already underway, but the significant difference is that unlike the disruption it often causes today, people will better learn to live with and adapt to this seamless world."

"It's happening. It is 9:39 p.m., and I am at home, but I am doing my work e-mail."

"I agree, already few people leave work at work. People regularly check and respond to e-mail at 10 p.m. The ability to disengage completely has nearly evaporated."

"Most people will discover that they want to get out of the house and be involved with peers."

"The boundaries between work and leisure have already blurred. I expect other social forces (e.g., traffic congestion) will encourage more telework (40% of a workweek), but I don't see it as a replacement for face-to-face social interactions."

> "As the basic organizing unit of human existence, I think it will be hard to change family dynamics. But could the diminished boundaries between work and leisure lead to more people choosing work that they love?"
>
> —Anonymous respondent

"For some; but not for all. For others, it will simply allow the sweatshop to be moved into the home, and at decreased costs to the corporation."

"No question about this. Major spike coming in 'Internet widows/widowers.'"

"The BlackBerry is the first wave of this...people are already working in meetings, hearings, church, wherever...After enough of this, the expectation to show up in the office will diminish as long as you give good e-mail."

"First of all, the effects have already happened. Second, telework may not expand as much as technology will allow, because there are advantages to working in physical proximity to others."

"I see a growth in work-in-cars as a means of blurring work/commute and home. It may be that carpooling happens because it provides people a way to work on the way to work."

"The boundaries between work and leisure have already diminished significantly, and I'm afraid that work is winning out."

"It's hard to ever be 'home.' No excuse for being offline."

"Homeschooling does not necessarily equate with technological uptake, though initial studies suggest that homeschooling families have a slight edge on nonhomeschooling families with respect to technology adoption. However, the loosening of the traditional workplace will allow for greater freedom for families. I'm not sure I would say 'sharply alter' family dynamics but alter, yes."

"I think that's already happened to a startling degree and that backlash and demand for personal/leisure time will result in mass turning-off of work connectivity at home."

> **"The trend is clearly for the rest of the workforce to join the new lifestyle. Make no mistake: This means more work time and less leisure time. For people with interesting jobs, this brings a more fulfilling life. For people with jobs that seem dull to them, this brings more cubicle misery —only now at home, too."**
>
> —Anonymous respondent

"First thing my sweetheart does in the morning is check his e-mail on a BlackBerry in the bed next to me. I suspect, though maybe wrongly, that there would be more cuddling were there no BlackBerries."

"It's already happened in my life and my family. Not just the Internet, but mobile communication ([e.g.,] phones, messaging) and other technologies for ubiquitous connectivity have this effect. It's going to be a bumpy road!"

"The boundaries have already diminished significantly. I think this will have a profound impact on the life of children."

"It's already a problem and it'll only worsen. P.S., I hate my husband's BlackBerry."

"I agree, but I don't think it will be a negative, just a reordering. For example, the traditional thinking of work Monday through Friday, 8 to 5, which is already largely a thing of the past, will continue to erode. But, the ability to work anywhere at anytime might in fact allow families to live where they choose without regard to proximity to work, which would allow for more quality time to be spent with families. It's already begun and I don't think it will stop."

> "I'm looking forward to seeing how this affects how Americans value work. I'm looking forward to a culture that values leisure and family time equally, and I think this could be a valuable stepping-stone."
>
> —Anonymous respondent

"As someone who works at home, I can say that this is already happening. It's a real challenge to segment work time from family time in an 'always-on' environment. Ultimately, this creates the same level of unhealthy distraction as ubiquitous usage of cell phones. We work too much in this profit-obsessed country, and unless employers start to trust their remote, home-based employees and don't constantly check up on them or try to force them into regular hours, and then demand last-minute work to be done at night or on the weekend, the telecommuters of the world might all be early heart attack victims."

"I agree that family dynamics may be altered for the digital elite. But most people, I suspect, will still go to work and use computers sparingly. Many schools still have a 50% dropout rate. Computers will not change that dramatically."

"The industrial model of home and work divisions is breaking down, and the cultural lag in recognizing this is creating a fair amount of stress already. Our current city/suburb infrastructure is predicated on the traditional division between a home place and a workplace, and technology is bringing about change faster than cities and suburbs can reorganize their infrastructure. This means that those individuals in families who must contend with these changes have to adapt much more quickly without the benefit of previous generations' experience and guidance. They are being forced into the role of teleworking pioneers. This will get sorted out by 2024 or so, and by 2034 people will not think twice about it. But we will see a strong generational gap, just like the gap when the farmworkers moved into the factory, or post WWII."

"As telework expands and invades the personal space, people will revolt and begin to place boundaries. We are already seeing this trend to some extent amongst the digital elite who are beginning to appreciate 'unconnectedness.'"

"I agree, but I don't like it. It's hard enough carving out the hours for family time (leisure); you're suggesting that everything will overlap. Everyone will have their own schedule, so I guess quality family time would be like scheduling a meeting with others that have busy schedules."

"On this summer's family vacation, I did work conference calls on my cell phone while my kids watched a video with headphones in the back of the van. You can call that a vacation, but it sure didn't feel like one on some days."

"You've got the causes wrong, but the outcome correct. The causes are wireless personal digital assistants such as BlackBerry and the Net itself, not telework and homeschooling, neither of which has really caught on."

> "Yes, it's already the case. My daughter and I even send e-mails to each other while we are in the same house; she is wirelessly connected to her laptop and I am on a desktop computer. We like the asynchronous convenience of this. We chose to answer or not and have a brief exchange, remind each other of details, set up rides to events. I find myself losing touch with friends who do not have e-mail."
>
> —Anonymous respondent

"There is a rise among Gen Xers to be home more with their kids. I think this will put increasing pressure on employers to allow telework and alternative schedules. If employers do not respond, I think we will see continued growth in self-employment and home-based businesses."

"I think this will happen sooner, and in 10 years we already see a counter movement. People are fed up by working all the time, or being interrupted by work during leisure time, so will claim back the free time and divide between work and leisure. Of course, there are always people that make work out of their hobby and think their work is their hobby."

"The difference between work and leisure will continue to exist. Family dynamics are already altered for many given the pervasive use of the Internet for chat, use of computers for games, and DVDs for movies. Working at home has been around for a while. The federal government says it wants more employees to work at home, but the reality is that the increase in productivity expected has not materialized. In private industry, some sectors may increase work at home, but others will not and probably cannot given the nature of their work—be it services or manufacturing."

"I imagine that we will no longer 'clock in and clock out' of work. It will be easier for families to schedule around personal needs, but it will also be easier for work to insinuate itself at home."

"This is either a utopian or an extraordinarily pessimistic prediction, depending on how you choose to read it. While both telework and homeschooling are increasingly enabled by online tools, that doesn't necessarily mean that there is huge pent-up demand by the broader society for either of these things. There are simply too many aspects of both work and school that require face-to-face contact. There are also special personal and interpersonal skills required to make either one of these things a success. It takes a special person to offer homeschooling to their child, or to work from home and maintain healthy relations with remote colleagues. The Internet doesn't change that. These things will grow, but likely at a modest rate for the foreseeable future."

Endnotes

1. See "The Broadband Difference," published by the Pew Internet & American Life Project, June 23, 2002, available at http://www.pewInternet.org/PPF/r/63/report_display.asp.

2. See "The Digital Disconnect," published by the Pew Internet & American Life Project, August 14, 2002, available at http://www.pewInternet.org/PPF/r/67/report_display.asp.
3. The phrase "knowledge worker" has been credited to Peter Drucker, who spoke on this topic in 1994: http://www.ksg.harvard.edu/ifactory/ksgpress/www/ksg_news/transcripts/drucklec.htm.

Part 11

Extreme Communities

PREDICTION: *Groups of zealots in politics, in religion, and in groups advocating violence will solidify, and their numbers will increase by 2014 as tight personal networks flourish online.*

Experts' Reactions	
Agree	48%
Disagree	22%
Challenge	11%
Did not respond	19%

Note. Since results are based on a nonrandom sample, a margin of error cannot be computed.

Many respondents pointed out that the Internet is like a pen, a microphone, or a telephone—all are wielded by good guys and bad guys. There is nothing inherent in the technology that gives advantage to one side or the others. Thus, many agreed that zealot groups will have an easier time sustaining themselves. By the same token, the experts are confident that all manner of helpful groups will have an equally easy time.

One expert wrote, "The Internet is a medium, not a motivator. It is possible, however, that the relative anonymity of the Internet will allow people to voice notions that would not be tolerated in polite 'arms-reach' society, thus more vitriol could be expressed without fear of social opprobrium normally expected when meeting with others face-to-face. In this sense, the Internet is like graffiti, only it can be targeted to the right niche."

A few experts criticized the sweeping nature of the prediction. For example, one expert wrote, "So far, we have little empirical evidence to show that online communication has such adverse effects. (That is, it would be hard to show that the Internet has had an isolated influence on bigoted actions among people who wouldn't have otherwise gone down that path anyway.) The jury is still out on how much the Net fragments people into little communities of people who completely agree with them." Fred Hapgood, a professional science and technology writer, had an even more pointed assessment: "These questions are very poorly formulated. Do you mean will the number of zealots increase or just the number of groups? I think the answer is no to the first and yes to the second."

> "Yes, but not only groups of zealots advocating violence. Also groups of 'zealots' advocating peace and nonviolent activism."
>
> —Noshir Contractor,
> University of Illinois
> at Urbana-Champaign

Another critic of the question gave it a more positive spin: "This is an interesting prediction. I tend to agree with it because the Internet, having broken boundaries of geography and linear time, enables niche groups to reach a 'critical mass' much more quickly and conveniently than in previous generations. But, of course, the zealots are not limited to religion and politics—they also include the quilters and the *Star Wars* fans and the peaceniks. The Internet itself is agnostic, and so should be your question."

Susan Crawford, a policy analyst for the Center for Democracy & Technology and a Fellow with the Yale Law School Information Society Project agreed, writing, "Although I think guilds will form, I'm not convinced that bad-guy guilds will be any more prevalent than good-guy guilds. People are generally nice to each other. Sure, like-minded people will find each other, but I don't think that's reason to adopt the negative language of this prediction. Yes, more groups will form. But this is a very diverse world, and there will be all kinds of groups."

A small group of respondents suggested that the Internet's positive force will outweigh any negative inclinations among its users. Robert Lunn of FocalPoint Analytics, a senior researcher, 2004 USC Digital Future Project, wrote, "I believe that personal networks will allow some undesirable groups to enhance communications among themselves and perhaps to even broaden their recruitment efforts. However, I also believe that enhanced communications and access to information is on the evolutionary path to freedom."

Additional Responses

Many other survey respondents shared quality remarks tied to the question about the Internet and extreme communities. Among them:

> "This depends on world politics. Certainly, they will try to grow and organize online as they are doing now—it will set up the battle between civil liberties and national security and which will win out is not clear at this time. If surveillance becomes better, this element may be rolled up more efficiently or eschew these networks for caves and smoke signals instead." —**Jonathan Peizer**, CTO, Open Society Institute

> "Communities can come together on specific issues and then disperse—more often than the case of large communities arising and staying together on fringe issues. Could there be more of this than today? Perhaps. But they will most likely stay small, isolated, and capable of only the type of occasional impact that we see have seen in the past." —**Benjamin M. Compaine**, a communications policy expert and consultant for the MIT Program on Internet and Telecoms Convergence

> "Tight personal networks will flourish, but governments will get to grips with how to monitor and control domestic groups.

Thus, they will publish their rot, but before they can press the go button on any dangerous activity, they will find themselves penetrated and cleaned up. International groups will be more difficult, but even here, increasing cooperation between nations will do much to control them. See the eEurope commitments on security." —**Steve Coppins**, broadband manager, South East England Development Agency, Siemens

"[I agree] only to some degree. In the case of any extreme group, IT can help with communication and coordination, but it does not completely replace in-person activities of a positive or negative kind. Yes, they will be able to organize better; no, they will not necessarily flourish as a result—that will depend on other factors." —**Ezra Miller**, Ibex Consulting, Ottawa, Canada

"The Internet is a boon for all kinds of groups that operate on the fringe of our society. Such groups were underrepresented in broadcast media and if anything, overrepresented in digital media. They are quick to adopt technologies that allow public outreach—which may result in some mainstreaming of their ideas in both senses of the term (the ideas will be more accepted by the mainstream, and the ideas will become more mainstream in order to be accepted). It also allows underground groups to maintain contact as they move across the globe." —**Henry Jenkins**, MIT Comparative Media Studies, author, *Convergence Culture*

> I disagree on 'tight personal networks.' These will be propaganda networks between people associated by fear, resentment, frustrations, but not personal relations."
>
> —Louis Pouzin, Internet pioneer, inventor of "Datagram" networking

"The Internet has seen a proliferation of self-reinforcing cyber-ghettoes of all types." —**Philip Virgo**, secretary general, EURIM

"The exact opposite will happen. The Web will be used by rational advocacy groups to expose the zealots for what they are. I am just hoping that people are by and large rational." —**Michael Wollowski**, Rose-Hulman Institute of Technology

"Nutcases that would normally be isolated all over the world can now meet in real-time on the Internet. This will become much more extreme in the next decade. The likely outcomes are not good." —**Robert Lessman**, owner, Quality GxP Inc., a consulting firm

"I guardedly disagree here. I think we will see more active and effective online groups, but the form these online groups take depends heavily on other social factors." —**Alex Halavais**, State University of New York at Buffalo

"The Internet will help weaken these groups." —**Bob Metcalfe**, Polaris Venture Partners, inventor of Ethernet and founder of 3Com

"I agree; however, as I said previously, I think law enforcement will have increased liberties online. As a result, it will become increasingly dangerous to express one's political and (extremist or non-Christian) views online." —**Lois C. Ambash**, Metaforix Inc.

"Personal networks have flourished offline spectacularly, so this can only become easier as global digital participation increases. But more moderate groups will also flourish." —**Bornali Halder**, World Development Movement

"Yes, but it will be balanced by increased organization by moderating influences, so they will cancel out." —**Paul M. A. Baker**, Georgia Centers for Advanced Telecommunications Technology

> "Groups of every kind will solidify and unite. The [provided] statement applies to stamp collectors, bee keepers, left-handed, one-eyed, hermaphrodite Albanian midgets."
>
> —Rebecca Lieb, Jupitermedia

"Individuals that would not otherwise interact will be able to more easily find and interact with compatriots. This will result in an increase in the unpredictability of small, organized actions, from violence by governments, individuals, religions, political, and other groups." —**Dan Ness**, MetaFacts

"People will turn to the Internet and be led like sheep on how and what to think." —**Tom Egelhoff**, smalltownmarketing.com

"If current trends continue, then fragmentation of the public sphere is one of the biggest challenges that a democratic polity must face." —**Albrecht Hofheinz**, University of Oslo

"The level of political discourse should rise in proportion to the penetration of the Net and the availability of trusted sources." —B. Keith Fulton, vice president, strategic alliances, Verizon Communications

"Thomas Jefferson said, 'From time to time, the tree of liberty must be watered with the blood of tyrants and patriots.' As our nation moves steadily away from the ideals on which it was founded, some radicals may indeed move toward violent resolution of their concerns. It stands to reason that the Internet will play a role in this. However, identifying serious religious or political adherents as 'zealots' is unfair demagoguery. Perhaps the question ought not to group religious and political zealots only with violent groups, but also those who advocate peace or more sustainable agriculture." —Daniel Weiss, Focus on the Family (Christian ministry)

> "I disagree. This ignores another countervailing aspect, which is the ease of access to information and opinion. It will be harder to isolate and brainwash initiates, which is always the tactic of such groups."
>
> —Mike Weisman, Seattle attorney, activist in the advocacy groups Reclaim the Media and Computer Professionals for Social Responsibility

"I do think the number of 'communities of interest' will increase, and they will become more tightly knit and integrated. More of these groups will be positively focused, but there will be negative elements as well. There will be good outcomes from this—it will be easier to learn about and understand issues from different points of view, which will enable people to make informed decisions about political and social issues." —Lyle Kantrovich, usability design expert, Cargill, blogger

"The political centre will get stronger, but so will extreme groups. The centre's ability to sustain itself will depend on its ability to respond to the outliers." —William Stewart, LivingInternet.com

"The power of the network to bring people together for both good and bad is possible. However, I believe good will always prevail." —Tiffany Shlain, founder, the Webby Awards

Anonymous Comments

The following excellent contributions to this discussion of the Internet and extreme communities are from predictors who chose to remain anonymous.

"This is already happening." —*Answer given by a number of anonymous survey participants*

"The decline of broadcast and the rise of narrowcasting via Web sites and blogs will support the flourishing of extremist groups."

"This has already happened in many cases. I suspect since this poses a patent threat to society (especially the violence) that countermeans will develop to mitigate the tendency."

"Electronic communities allow smaller, more fringe groups to form and sustain themselves."

"I believe that more security measures will be in place that will be able to monitor these types of networks and restrict/prevent use for violent means."

"These will be balanced by better communication and trust across other types of groups."

> "Groups such as these will recognize that the Internet only increases their visibility. While they continue to use the Internet for publicity purposes, most of their activities are likely to remain off the Internet—and out of sight."
>
> —Anonymous respondent

"It's simply hard to know... the strength of these groups is likely independent of technology and will depend more on the political climate, etc."

"Agree with the prediction, but not with the online nature. I think these will continue to be driven by geographic or cultural affinities, with online tools possibly enhancing already strong social ties, but not as an essential driver."

"These groups depend on secrecy. The Internet will tend to expose these groups rather than hide them."

"This is a very dangerous phenomenon—groups of people reinforcing each other's beliefs, and narrowing the range of information to which they're exposed."

"The number of fringe groups may increase, but their membership will be small and their impact will be limited. Indeed, the behavior of 'insurgents' now in Iraq seems to indicate this."

"I'm not sure that the numbers of these groups will actually increase. But then we don't need a lot of such groups to be worried. A few really bad ones will do, and that will probably happen."

"Generally, interest groups are well served by the Internet, and tiny communities can become small communities more easily. They will be (are) global."

"They will solidify and increase, but so will governments' abilities to track and extinguish them. This will be an ongoing ebb and flow."

"Like civic groups, these organizations are too small to have the drive and funding to develop comprehensive network infrastructures."

"Smart zealots will avoid the Net…it will lead the authorities right to them."

"It is possible that Internet-based extremist groups will be more visible than their non-Web versions, and hence, generate countermovements."

"Groups of zealots appear to be on the rise in all areas. Whether this leads to violence, I don't know. But it is perhaps one of the sharpest dividing times in my lifetime. It is like we are in a civil war, with brother against brother and neighbor against neighbor. There is no common good in the USA or the world at the moment. Technology and online-ness are minor facilitators in this process. Tight personal networks online are unlikely but online as a filter of information, with only trusted sources, is very likely. 'Trusted' networks might be a better way to talk about online groups. I 'trust' them for the truth about America's involvement in Iraq (whether the trusted source is Fox News, the Mormon Church, or al-Jazeera doesn't really matter, as long as it is your trusted group)."

"These are two different questions: Will fundamentalism expand? Does the Internet lead to tight personal networks? As to the second, no, I do not believe that to be true. The Internet leads to diversity and viewpoint exchange. It leads to access to unheard voices. It leads to dialogue. But the first question is: Will alienation expand? I fear it will. Capitalism now runs unchecked by communism. The gap between the rich and the poor expands. The rich in the U.S. make more than other countries. Violence and fear are making imbeciles out of people, and the lofty goals of the U.S. democratic experiment have all but been forgotten...Karl Marx may yet be right—that capitalism will collapse in revolution of some type and be transformed into something else. Simply because the USSR failed does not mean that Marx was not right that capitalism has within it the seed of its own destruction."

> "Information and technology should become a healer, not a destructive force. History has a way of repeating itself rather than being significantly modified by technology."
>
> —Anonymous respondent

"Well, they're there now, but I don't foresee a marked increase of online presence—at least for groups advocating violence and illegal activity—because I don't think the security will be there: The digital world is very sticky, and their actions will leave traces."

Part 12

Politics

Prediction: *By 2014, most people will use the Internet in a way that filters out information that challenges their viewpoints on political and social issues. This will further polarize political discourse and make it difficult or impossible to develop meaningful consensus on public problems.*

Experts' Reactions	
Agree	32%
Disagree	37%
Challenge	13%
Did not respond	18%

Note. Since results are based on a nonrandom sample, a margin of error cannot be computed.

An October 2004 report from the Pew Internet & American Life Project shows that even as political deliberation seems increasingly partisan and people may be tempted to ignore arguments at odds with their views, Internet users are not insulating themselves in information echo chambers.[1]

Most experts disagreed with the prediction or challenged its premise, arguing that there are many other forces at work in political discourse. For example, Jorge Reina Schement, director of the Institute for Information Policy at Penn State University, wrote, "It is not the Internet that drives polarization of the electorate. Rather, polarization stems from the inability to find common ground when values differ. Polarization in American society will continue; the Internet will serve to abet this tendency."

Another expert cited historical precedent, writing, "Most people use new media so they know what is important, what has happened and what may happen. The partisan press isn't new—this nation was born out of a partisan press. And it certainly has been the norm in other parts of the world for many, many years. The bigger concern is the concentration of ownership by large corporations."

B. Keith Fulton, vice president for strategic alliances at Verizon Communications, wrote, "The Net should have the opposite effect on 'most people.' Sure, crazy folk will find crazy folk. But the masses will use the Net for their first news and will go to trusted sites for affirmation and/or information that they seek."

One expert grew tired of the survey mixed-scenario-style format and spoke for many when he wrote, "It's very difficult to know how to answer these questions, since they are usually of the form, first A will happen and that will cause B, which will cause C. Suppose I think A and B will happen, but not C?? This question is just one example among many. Besides, what is 'meaningful consensus'? Did we ever have a 'meaningful consensus' at any point during the 19th century, when there was no Internet to speak of?"

> "There will always be people who want to be challenged by opposing points of view, and as long as those points of view can still publish, the Internet will make it more possible to access them with ease."
>
> —Noshir Contractor,
> University of Illinois
> at Urbana-Champaign

ADDITIONAL CREDITED RESPONSES

Many other survey respondents shared incisive, overarching remarks to the question about the Internet and political discourse. Among them:

> "Thank heavens for crosscutting cleavages! Yes, people may spend more time networking with fellow Republicans/Democrats/environmentalists/fundamentalist Christians. But they'll also spend more time networking with fellow Red Sox fans/Labrador owners/amateur carpenters/Edith Wharton fans. And anyone who has ever been part of an online community knows how hard it is to prevent off-topic threads and discussion. Politics will always pop up in...so it may be that there are MORE opportunities for bridging, as well as bonding online." —**Alexandra Samuel**, Harvard University, Cairns Project (New York Law School)

> "Because the Internet is such a gift economy, we'll continue to follow links from our friends. But we'll also develop meaningful shared spaces for discussion, and we'll be able to see one another there in the form of avatars. Both will happen. Consensus will be both easier and harder. Visualization of information will be the key development in the next 10 years, and it may help consensus emerge. But groups will be tighter."
> —**Susan Crawford**, a policy analyst for the Center for Democracy & Technology and a Fellow with the Yale Law School Information Society Project

> "I suspect that people will be able to effectively filter the information they are exposed to, but I also think that people today are capable of selectively perceiving the information they are presented. I suspect that the move toward polarization will be accompanied (and counterbalanced) by new forms of public deliberation and exchange." —**Alex Halavais**, State University of New York at Buffalo

> "The first part of the prediction has been established long before today—selective perception. At the start of the 20th century, competing newspapers had an acknowledged labor or a business or a political point of view, and it was typical for people to buy the paper that reinforced their viewpoint. Nothing new here. Make it 'impossible' to develop consensus?

Doubtful. Compromise will survive." —**Benjamin M. Compaine**, a communications policy expert and consultant for the MIT Program on Internet and Telecoms Convergence

"I think there is enough diversity and leakiness in conversations and personal networks that alternative viewpoints will still be realized." —**Barry Wellman**, University of Toronto

"Well, we already have that (polarized discourse) today, don't we?" —**Bill Eager**, Internet expert

"The only countervailing influence would be for major publications to make it difficult to receive totally customized feeds. That is, in order to receive *The New York Times'* editorial-page Web feed, a person would need to receive both David Brooks and Paul Krugman—all or nothing. We are already seeing the effects of polarization on cable TV viewership. During the Democratic convention, CNN viewership was up; during the Republican convention, Fox viewership was up. As the blogosphere grows, it will be possible to insulate oneself almost completely from opposing viewpoints. I would ideally like to see a voluntary 'pairing' of opposing online publications that would encourage people to expose themselves to alternative—even abhorrent—viewpoints. How possible is that?" —**Lois C. Ambash**, Metaforix Inc.

> "This is the way most people run their lives, Internet or no Internet. The Internet makes it just as easy to get a quick overview of the political landscape from all viewpoints as it does to filter out opposing views. If you prefer to have your thinking go unchallenged, you'll choose the Internet in the latter manner; if you seek to widen discourse, you'll use it to keep tabs on multiple viewpoints."
>
> —Rose Vines,
> freelance technology journalist,
> *Australian PC User,*
> *Sydney Morning Herald*

"There will be so much information that people will deal with it by filtering. It will be possible to get whatever viewpoint is desired, and it will be favored." —**Ted Eytan**, MD, Group Health Cooperative

"Means of integrating mass media communications will evolve to make sure we keep a good mix of materials. If you don't want to know the news you don't read the paper or watch TV—for some this will be true online as well. But there will always be those who want to lead and together build consensus for dealing with the problems of the day. Just because we have the Internet, it doesn't mean we will lose the silent masses out there." —**Liz Rykert**, Meta Strategies Inc., Toronto, Canada

"A subclass that spans political and ideological divides may arise, one that encourages people to question what they're being told." —**Mack Reed**, Digital Government Research Center, USC

"This might be true for some, but many are using and will use the Internet as an easy and harmless way to explore differing points of view. Listservs and online chat provide opportunities to interact and see 'conventional wisdom' challenged. With newspapers publishing online, there will be greater accessibility to opinion and greater opportunity to interact with opinion writers. It is difficult to come to the conclusion that the Internet will polarize political discourse to the extent suggested." —**Ezra Miller**, Ibex Consulting, Ottawa, Canada

> "If people just want to hear their own point of view, they don't need to pay for an Internet connection—they can just listen to themselves and their like-minded friends at no cost."
>
> —Elliot Chabot,
> senior systems analyst,
> House Information Resources,
> U.S. House of Representatives

"If the right new services emerge, this won't turn out to be the case. And I have faith that the right services will emerge." —**Gordon Strause**, Judy's Book, a social-networking Web site

"People welcome controlled dissent—they want to know that there is another perspective out there, and many pursue the opportunity to challenge that perspective. Technology and the Internet are facilitators for that." —**Brian Reich**, Internet strategist for Mindshare Interactive and editor of the political blog campaignwebreview.com

"Economic, social, and political developments, certainly in the U.K., suggest the opposite. Individuals are becoming less partisan and less certain in their views. Therefore, we may have citizens with a mishmash of views. What might be changing (and not just because of technology) is the very nature of politics." —**Nigel Jackson**, Bournemouth University, U.K.

"Frankly, I think people like a good fight over these issues. They can filter out these viewpoints now, and they don't. I am not afraid of this." —**Arlene Morgan**, Columbia Graduate School of Journalism

"To some degree this is already happening...Cass Sunstein is at least partially right. But as I have argued in the *Boston Review*, I think he is underestimating the range of different kinds of affiliations people have and the degree to which social, cultural, and recreational connections may be more important to them than political alliances. The result is that many alternative perspectives will get through such filters because they will be part of other kinds of conversations people are holding." —**Henry Jenkins**, MIT Comparative Media Studies, author, *Convergence Culture*

> "I can't see how this would be possible...the Internet is the ultimate free-speech printing press."
> —Graham Lovelace, Lovelacemedia Ltd., U.K.

"I disagree with the 'polarizing' statement. One of the major benefits of the Internet is access to a variety of different points of view and sources of information. I think this will continue to be a benefit of online communication." —**Gary Kreps**, George Mason University, National Cancer Institute

"As Internet users increasingly set filters and personalize, there will be a backlash fueling the rise in services and sites devoted to pure serendipity. Many people will gravitate away from detailed personalization in favor of the pleasure of not knowing what's next. Those who read a daily newspaper, even though the same content is on the Internet for free, often do so because they enjoy not knowing what's on the next page. Editors will rise in importance, as people realize that too much personalization and individual search stifles creativity and curiosity. At the broadest level, there are exactly two ways to use and deploy content on the Web. Most organizations put too much effort on one just way: 'Answer my question.' While not spending enough energy on the other: 'Tell me something.' Too often the content that's deployed on sites helps visitors solve only half of their needs. The more obvious way to conceive of Web site navigational design is to help users answer a question. To illustrate this concept, consider one of the Web's best-known sites, Google, which in its purest form exists only to answer questions. With a site or content product organized only around answering questions, users must already know what they want before proceeding. But people also need services or sites to tell them something. Contrast Google with another famous site, *Drudge Report*. It doesn't answer questions at all; rather, it tells visitors stuff they didn't think to ask. Organizations will build sites to encourage serendipity and browsing." —**David M. Scott**, Freshspot Marketing, *EContent* magazine

> "It's already happening when people buying one book on Amazon are given suggestions as to what also to buy, and the subjects are always related. Similarly, Web sites and blogs attract like-minded people and rarely link out to sites/blogs with a different viewpoint."
>
> —Bornali Halder, World Development Movement

"Increased communication is unlikely to produce narrower communication. Even those groups against the mainstream generally discuss it, if only to denounce it. More information will always be only a click way." —**William Stewart**, LivingInternet.com

"The Internet, as is true of cable or satellite TV and print publications, will continue to include a mix of the highly specific and the more general. People will still have broader views available to them in the media they consume." —**John B. Mahaffie**, cofounder, Leading Futurists LLC

"Technology doesn't polarize people; personal, professional, and organizational conduct polarizes people. The Internet will not substantially affect the trend of hyper-hysteria already at work today in politics. If anything, the Internet will continue to provide alternative voices and outlets that allow for a greater dissemination of ideas beyond the increasingly radical and liberal mass media. I think the lack of visionary leadership and personal character is more responsible for our nation to find meaningful consensus. No technology can cause or change that." —**Daniel Weiss**, Focus on the Family (Christian ministry)

"In fact, although we like to read information from people who agree with us, the Internet makes it even easier for us to seek out and read opposing views." —**Mike Weisman**, Seattle attorney, activist in the advocacy groups Reclaim the Media and Computer Professionals for Social Responsibility

"It will be easier to learn about and understand issues from different points of view, which will enable people to make informed decisions about political and social issues. This prediction also has the misconception that we 'develop meaningful consensus on public problems' today. People will choose when they want to filter out dissenting opinions and when they want to understand other viewpoints." —**Lyle Kantrovich**, usability design expert, Cargill, blogger

> "People will be increasingly exposed to alternate viewpoints. My personal experience is that people are engaging in constructive, spirited dialogue far more now than ever before as a result of the Internet. Plus we are better informed and can check our facts quickly and easily."
>
> —Peter W. Van Ness, principal, Van Ness Group

"By 2014, people will acquire, through interactive technologies, the ability to filter most information they are exposed to, not just that which arrives through Internet means. I'm not sure this on its own will change anything. There is little difference in the societal outcome in filtering by choice and being restricted from exposure certain types of information by mass media ownership concentration. It returns to the notion that there are two types of information consumers—those who actively seek it, and those who are passively subjected to it. Consensus building has more to do with promoting societal principles of participation and the ideas that diversity of opinion, critical thinking, and open discussion are essential things to a healthy democracy. Those who learn that filtering is the best way to get along will filter more effectively through interactive choice. Those who learn that health, both mental and societal, comes from open discourse and respect for a diversity of opinion, will use the Internet as they do now, as a tool to seek information that in some instances can also be used enable dialogue." —**Sam Punnett**, FAD Research

"People do this already, in print media and television. Liberals don't watch Fox News. Overlap in the subscriber lists of the American Spectator and (insert liberal rag here) are small. Reasoned discourse will continue to flow, but it won't make headlines any more than it does now. Net result: no real increase in polarization traceable to the Internet specifically."
—**Mike O'Brien**, The Aerospace Corporation

"If current trends continue, then fragmentation of the public sphere is one of the biggest challenges that a democratic polity must face." —**Albrecht Hofheinz**, University of Oslo

"People will turn to the Internet and be led like sheep on how and what to think." —**Tom Egelhoff**, smalltownmarketing.com

"People filter all the time. Fox News vs. the BBC. Deleting some e-mail, reading other messages. This is just another channel that enables filtering. And exposing oneself to broader views. Depends who's doing it." —**Rebecca Lieb**, Jupitermedia

"I believe people truly seek accurate information, from whatever source. Filters will play a role, but most will want as much information from all viewpoints as possible." —**Ted Christensen**, coordinator, Arizona Regents University, overseeing development of e-learning at Arizona's three public universities

"Many people will still check out 'mainstream' media, and many will still ignore it and/or interpret it eccentrically. But that's always been the case. Centralized media (like the Big Three TV networks) probably provided more of an illusion of society-wide consensus to people near the center of power than they did the real thing. The only thing that may change is that people inside the beltway believed that Walter Cronkite represented the majority or the center, whereas they're now realizing that Dan Rather is just another guy with a point of view." —**Tom Streeter**, University of Vermont

Anonymous Comments

The following contributions to this discussion of the Internet and political discourse are from predictors who chose to remain anonymous.

"I still think there is a good chance that the center will hold. People have a lot in common with each other."

"Are you kidding? The Internet is the greatest thing that has ever happened to expand the number of voices that need to be heard."

"I don't know if it makes it impossible, but it has contributed greatly to the phenomenon of the divided nation."

"I believe the Internet will allow for exactly the opposite—less censorship and more diverse range of information sharing."

"Filtering happens in the print world, and it happens in the Internet world. Even if technology makes it easier…I do not think increased use of filtering will polarize discourse."

> **"I believe that people realize—or will realize—that diversity is needed for a good decision-making process. I think the notion that consensus, or common good, is the goal will give way to enlightened self-interest. That is all that can be expected, it also is all that is needed. The polarization is independent of the Internet."**
>
> —Anonymous respondent

"The Internet will amplify existing tendencies for both expansive and narrow viewpoints. The way we reach consensus will need to adapt."

"I may be an optimist, but I do observe that people of various convictions actually seek information. There are zealots, but they are not a majority."

"People tend to consume the media—whether the Web or print or TV—that agree with their personal point of view. I don't believe you can attribute the Web's filtering capabilities as the cause. When you choose any information, you filter."

"I do expect polarization to be increased by the expansion of more narrowly defined/targeted information flows."

"People prefer to operate in their cultural comfort zone ingrained from a very young age."

"People will lose the ability or desire to consider the potential validity of another point of view. The outcomes of this could destroy the ability of any democracy to function. Democracy demands a certain respect for the loyal opposition."

> "In the U.S., we will have more enlightened individuals—those who will have grown up using the Internet and being able to better discern truth from fiction, political rhetoric from fact."
>
> —Anonymous respondent

"Most people will not understand the bias and filtering capabilities of technology and will not be able to affirmatively choose filters. Information will still be filtered, but more likely in ways that are surreptitious and insidious."

"Here, the problem is with 'most people.' 'Most people' do filter information to reduce the quantity of information that challenges them and increases the flow of information that supports their world view. But I do not see that the Internet will exacerbate this trend."

"I hope not and, in fact, believe just the opposite."

"Human nature as it is, people are likely to use the technology to do this. Whether the second sentence comes true or not depends upon what other discourse takes place in their lives—television, the workplace, etc. I am not too concerned about this possibility."

"The large center of people is not polarized highly in its politics and is open to different perspectives. Polarized people of the right and left, though, are likely to become more polarized."

"People already do this with other media—what books and newspapers they read, the radio shows they listen to, the television shows they watch. The Internet is no different."

"Most people do not use even the most basic filters! Most people do not belong to what we Brits call the 'chattering classes,' but to the working classes, and the working classes derive their information from well-established channels, which set a definite bias, but do not create extreme opinions."

"I'm a progressive politically, but I wound up reading the Instapundit (prowar) blog every day, and supported the war. The Net lets you believe what you want to believe, and have it reinforced by fellow believers."

"Information overload, and invasion of the personal desktop, occurs more and more, and, yes, I believe we will filter. Every spam message, troll on a listserv, and commercial message invades our personal computer space. I think we will fight to have that back using whatever filtering we can. Will we cut out political and social discussion? How many people listen to them now? Those who filter now, will filter then. I don't see that debate will be any more or less polarized that before. It may just be more visible on Internet discussions."

"The second proposition doesn't necessarily follow from the first. I suspect that exposure to challenging views, as in the current mass media that tend to exaggerate them, has a polarizing effect. Ignorance could engender tolerance, conversely. I agree with the first statement, however."

"By 2014, most citizens will use the Internet to investigate issues on their own, leading perhaps to a unified base of facts on which to formulate their own judgments."

"I think most people will know more about different points of view than they did in 1994."

"Sunstein is wrong! Online discussion is not as polarized as what Wilhelm, Davis, etc., made out (see Stromer-Galley, Wright, etc.). What is needed is efficient online mediation—see Hansard Society, etc."

"This is always a popular prediction, but I don't see that happening any more than it does now: (a) because I'm not sure most people are so terribly frightened of views that differ from there; and (b) actually filtering in information on the basis of content is very difficult, especially if people set out to subvert said filtering."

"I agree with the first sentence. However, I think that television, and to a lesser extent print publications, will continue to play a major role in shaping consensus, as will social institutions such as clubs, political parties, and churches."

"Currently, people watch programs or read publications that meet their viewpoint but they are still aware of opposing issues. I don't think this dynamic will change."

> "I would like to think that the opposite will happen—that because information is so easy to get and accessing information becomes more private (the neighbors won't see you buying alternative press at the local shops) that people could actually access more diverse opinions. This probably won't happen. But I think that e-mail might help because people are often sending friends and family links to articles and information that the recipient would not necessarily seek out on their own."
>
> —Anonymous respondent

"Highly unlikely, in part because most Internet users won't be sophisticated and/or interested enough to set their preferences so narrowly."

"There will be freedom of speech on the Internet that cannot be suppressed in a free country. Alternative sources of information, Web bloggers, instant-messaging devices, and other Internet tools are too numerous to censor."

"Unfortunately, it does seem that people are signing up for 'e-mail alerts' and such from media that reflect their own biases. However, the general media (cable stations, newspapers, etc.) is also moving to putting a stake in the ground that pegs them publicly to a specifically 'right' or 'left,' liberal or conservative, orientation to reporting. Internet filters just exaggerate the effect of this."

"This is the way we have always used newspapers, television and radio, and mass media generally. Do we think the Internet is in some way different, or that its network effects amplify the problem?"

"I think this could but might not necessarily happen. We're already seeing fragmenting and segmenting in media usage and the way media is sold and distributed. Combined with there just being so much info out there, and ease of connecting with like-minded people, I can see many people finding it easiest to read or participate [in] what interests them, while tuning out of what doesn't."

"As the trends predicted by...George Orwell in *1984*, A. Huxley in *Brave New World*, and Herbert Marcuse in *One-Dimensional Man* continue to emerge, people will increasingly get sanitized propaganda-cum-news. Again, this is a function of social relations, not technology."

"People will increasingly encounter more diverse views by using the Internet."

"This accepts the Western bipolar view of the world, where issues are divided between the resolutions *A* or *Not A*. But of course, that is not the way the world works. Issues are dynamic and interrelated. Resolutions are complex. Shifts in small paradigms follow chaos theory and result in dramatic shifts in resolutions and situations. So great will the information exchange become that isolationism will be all but impossible. Why, some nit may ask a survey limited to predictions about the future, and get back answers about Karl Marx, Plato, and theories of society and alienation."

"A lot of people probably will, but I have enough faith in the American people to believe that they will want to occasionally encounter different viewpoints. I also think that since people use the Internet based on their personal interests—which may or may not include political and social issues—they will encounter different viewpoints in the course of their Internet use, even if they are filtering according to their interests. In other words, I don't think people's interests always fall along political lines."

Endnote

1. See "The Internet and Democratic Debate," published by the Pew Internet & American Life Project, October 27, 2004, available at http://www.pewInternet.org/PPF/r/141/report_display.asp.

PART 13

HEALTH SYSTEM CHANGE

PREDICTION: *In 10 years, the increasing use of online medical resources will yield substantial improvement in many of the pervasive problems now facing health care—including rising health care costs, poor customer service, the high prevalence of medical mistakes, malpractice concerns, and lack of access to medical care for many Americans.*

Experts' Reactions	
Agree	39%
Disagree	30%
Challenge	11%
Did not respond	19%

Note. Since results are based on a nonrandom sample, a margin of error cannot be computed.

A December 2002 Pew Internet & American Life Project survey showed that 80% of Internet users, or about 102 million Americans, have searched online for at least one of 16 major health topics.[1] An educated consumer stands a better chance of getting good treatment and the Internet can be a significant resource for that health education process. An increase in the use of Internet health resources is not a cure-all, however. For example, low health literacy limits many Americans' ability to understand the information available online.

Many experts pointed out that the Internet's strengths (information, communication) will have little effect on the U.S. health system's weaknesses (inequality of access to care, among others). Those who agreed with the prediction, however, were enthusiastic about the potential benefits of electronic medical records, remote monitoring of patients, and peer support.

Pamela Whitten, an associate professor at Michigan State University and a Senior Research Fellow for Michigan State's Institute of Healthcare Studies, wrote, "External barriers (such as legal/regulatory/reimbursement/organizational) still trump many efficiencies offered by the Internet."

One respondent wrote, "In fact, I think the increasing use of online resources will actually exacerbate most of these problems in the short term of the next decade, since medicine is still on the steep side of the adoption and learning curve in IT and use of the Internet in particular. The costs of the IT investment required by HIPAA[2] alone will add measurably to health insurance premiums, today and for at least the next 5 years. Part of the problem is that, even though IT and online improvements in these areas are likely to be beneficial to the very great majority of consumers, the potential cost of nagging problems or spectacular single failures is devastatingly high."

One expert wrote, "The rise in automation and self-serve options (such as we see in self-checkout lanes at supermarkets) is training our society not to rely on service from other human beings. My sense is that it is likely that patients will increasingly use the Internet to act as their own doctor, coming

> "Online technology will only codify current health policy that fragments care and underserves a significant minority of the American population. Real reform, including finance reform, is needed, which will result in cost reduction, facilitated by online access."
>
> —Ted Eytan, MD,
> Group Health Cooperative

to their physician with not only complaints but 'solutions,' the quality of which will be suspect."

Another wrote, "Both the rising health care costs and its result, the lack of access to medical care, are the results of health becoming a source of profit for investors who already have money. Nothing inherent in the Web fixes that. Ditto for poor customer service. There is a potential gain in the widening of access to medical specialties for those who have the money—digital records, easy transmittal of test results, and MRIs for second opinions or consultations, etc.—but the increasing corporate seizure of what had traditionally been a private, in-person matter between physician and patient also brings with it denial of benefits, and thus of services, that act as a counterweight to those benefits for far too many Americans, and world citizens."

On a more optimistic note, one expert who works in health care wrote, "We are already finding the Internet useful as a cheap way to distribute life-saving or promoting information and services to far-flung areas. Believe that there may be some (not a lot) of savings that can be put into other programs." Gary Kreps, a professor at George Mason University and formerly the chief of the Health Communication and Informatics Research Branch at the National Cancer Institute, wrote, "With increased access to relevant health information, better decisions will be made by health care consumers and providers."

Many respondents wrote that they both agreed and disagreed with the statements (one complained that an "unreasonable connection" between two predictive statements was growing tiresome). For example, one expert wrote, "It will be very helpful, but not for the reasons listed...The main advantage will be for peer support and information sharing."

ADDITIONAL CREDITED RESPONSES

Many other survey respondents shared incisive, overarching remarks to the question about the Internet and the health system. Among them:

> "Online medical resources could certainly have a substantial impact on health care, particularly in diagnosis, consultation, [and] coordination. Certain types of mistakes could be

lessened. But it is not likely to have a substantial impact on health costs, coverage, malpractice suits, or access. Those needs require other avenues to remedy." —**Benjamin M. Compaine**, a communications policy expert and consultant for the MIT Program on Internet and Telecoms Convergence

"Not a chance. The financing mechanism guarantees higher costs." —**Fred Hapgood**, Output Ltd., technology writer

"The Internet could fill a huge vacuum in health care—the institutions are already weakened, the needs are critical, the demands are increasing—the time is ripe." —**Christine Geith**, Michigan State University

"A series of structural changes in the insurance and health care industries must be made to break down the barriers to making such online resources truly useful. Whether or not the Congress and the interested industries have the courage to make those changes is another question." —**Bradford C. Brown**, National Center for Technology and Law

"This may take more than 10 years to accomplish, but it will happen. Much of our interaction with doctors involves talking to them—describing symptoms and events, etc.—and they give us advice, assurance, and guidance and prescriptions. That doesn't have to happen in a physical office. This will all start with standardized, digital medical records, easily accessible by medical personnel. Then e-mail exchanges between a doctor and patient, including routine prescriptions, scheduling, referrals, and follow-up. Eventually, real-time videoconferencing (like Apple's iChat) will permit 'virtual' office visits, coupled in some cases with remote-sensing monitoring of vital functions such a blood pressure, pulse, temperature, etc. In the longer term, the Internet will be able to keep more of the elderly out of nursing homes for a longer time: At-home

> "Health care in the United States is likely to get progressively more expensive, less affordable, less available to those who need it, and more plagued with mistakes."
>
> —Gary Chapman,
> LBJ School of Public Affairs,
> University of Texas at Austin

patient care will be made possible by remote monitoring...If truly 'big broadband' gets into enough homes, at the right price, it can enable enormous savings in health/nursing home costs for society as the baby boom generation ages." —**Gary Bachula**, Internet2

"The benefits of any technological advances may be negated by timeless constants. While some surgical techniques may be improved or automated, and less educated doctors may learn better treatment protocols online, the bulk of the work still comes down to doctors' ability to spend adequate time with patients and use the best medicines, technology, and treatment at hand. The rise of malpractice has turned it into a less lucrative profession, which may limit the numbers of new doctors certified each year." —**Mack Reed**, Digital Government Research Center, USC

> "The entrenched silos of health care are huge, and the Internet alone cannot take this on, even as people get empowered with better medical knowledge."
>
> —Barry Wellman, University of Toronto

"Access to medical information is a good thing and is already happening. But there are strong forces afoot to suppress information or access to non-FDA approved services. For example, there are bills in Congress to remove all vitamin supplements from the marketplace and make limited amounts of supplements available by prescription only. If this happens, people's medical choices will decrease and chronic illness will increase because people won't be able to maintain healthy preventative practices." —**Peter Denning**, Naval Postgraduate School, Monterey, California, columnist, *Communications of the ACM*

"Too many people will use the Internet as fodder for becoming more demanding, more expensive health care consumers." —**Alexandra Samuel**, Harvard University, Cairns Project (New York Law School)

"I see a growth of misinformation on the Internet and the problem of people getting conflicting information regarding medical conditions not fully understood or on which physicians and scientists differ. People will be more informed about medical conditions but not necessarily better informed or better able to make decisions. Medicine is not a good place for autodidacts." —**Stanley Chodorow**, University of California–San Diego, Council on Library and Information Sciences

"Electronic medical treatment has been around for more than a decade. Without the investment of huge amounts for remote diagnostic equipment and video networks, it's unlikely that there would be much real change in care. More information is and will be available, but actual treatment is declining overall." —**Mike Botein**, Media Center, New York Law School

> "Online medical resources will have next to no effect on the delivery of adequate medical care to all. A political solution is required, and throwing technology at the population won't fix a fundamentally flawed system."
>
> —Rose Vines,
> freelance technology journalist,
> *Australian PC User,*
> *Sydney Morning Herald*

"I think this is essentially true, but I am not convinced that technology in health care will either decrease costs or increase access to care for the disadvantaged. I think technology will go a long way toward improving the care and the level of service that is delivered, and I have hopes for President Bush's EHR initiative. But I am not yet convinced, because of the costs of building and maintaining computerized networks and the software and devices needed to run on it, that the costs themselves will go anywhere but up. And as more people lose insurance, this will make care even more difficult to obtain, IT or no IT." —**Kevin Featherly**, news editor, *Healthcare Informatics*, McGraw-Hill Healthcare Publishing

"The power of information technology and the severe challenges to the health care system will drive change." —**Howard Rheingold**, Internet sociologist, writer, speaker

"This is very much a moving target. As medicine itself advances at an unprecedented speed, the demands on the system are likely to continue to outstrip their adequate provision."
—**Alex Halavais**, State University of New York at Buffalo

Health System Change

"I expect improvements, but not necessarily dramatic ones here. The privacy issues will be a major obstacle until (and even after) widespread authentication exists." —**Terry Pittman**, America Online, Broadband Division

"Improvements will be driven by HMOs and insurers. Improvements in customer service, access, and similar patient-centered concerns—to the extent they occur—will be fortunate byproducts rather than primary concerns. Some dramatic improvements that could readily be achieved—such as online consultations for common ailments with physicians who have not previously met the patient in person—will be forestalled or eliminated by physician-interest groups." —**Lois C. Ambash**, Metforix Inc.

> "This is far too optimistic. Some improvement will ensue, particularly in the areas of customer service, information sharing, and eventually with remote surgery (i.e., one seasoned surgeon walking a remote, less experienced surgeon through a procedure live). However, serious attention needs to be paid to privacy concerns now, or a backlash of mistrust could prevent people from going to doctors for fear of having their illness made public."
>
> —Peter W. Van Ness, principal, Van Ness Group

"Medical mistakes and malpractice will still happen. People will have access to more information to challenge decisions. The Internet will also allow better research of doctor records to prevent it. Certainly, administration functions may be better dealt with online. Information is half the medical challenge and the Internet will help facilitate its dissemination more efficiently." —**Jonathan Peizer**, CTO, Open Society Institute

"So long as the poor, the old, and the minorities are less connected than the wealthy and educated, we cannot expect to dramatically influence the health system, whose highest users are from the same disadvantaged groups." —**Tobey Dichter**, Generations on Line, nonprofit Internet-literacy agency

"Telemedicine is already having a significant impact on remote communities. Streamlined insurance processing, elimination of prescription mistakes, [and] better and confidential accessibility to records including longer-term archiving (medical imaging, test results) are all on the way to improving health care—though there are still hurdles to overcome (e.g., getting the insurance companies to cooperate)." —**Ezra Miller**, Ibex Consulting, Ottawa, Canada

"I can see the Internet perhaps improving customer service and some rural access to medical expertise. But I can't see how this will lead to lower health care costs, less medical mistakes, or more medical care for Americans. These problems are systemic and not related to technology. They are political."
—**Mark Glaser**, *Online Journalism Review*, Online Publishers Association

"The Agency for Healthcare Research and Quality has put a lot of funding toward research in health information technology with the goal of improving patient safety. While technology is not a panacea for eliminating medical errors, it can automate some error-prone processes. Recent research shows that it can introduce previously unheard of errors as well, so caution is warranted. Rising costs could be slowed and customer service improved by the efficiency of online medical resources. Access to health care through long distance consultations with specialists will increase as telemedicine becomes e-medicine."
—**Elizabeth W. Staton**, University of Colorado at Denver Health Sciences Center

> "The transformation of medicine is coming—the dire situation we have now will make it inevitable. It will be a tough paradigm shift to make, but new generations of medical practitioners will bring it with them."
>
> —John B. Mahaffie,
> cofounder,
> Leading Futurists LLC

"Most developed countries are aging fast. Having more online resources will help them but can't prevent the actual care that has to be given to a growing number of elderly, and, therefore, costs will still rise. I do think the people will be better informed, so the doctors have to give their best care possible before they 'shop' somewhere else." —**Egon Verharen**, innovation manager, SURFnet (Dutch national education and research network)

"Such systems will be vulnerable to abuse (hacking, security, and privacy invasions)." —**Bornali Halder**, World Development Movement

"It's not the ability for patients to look up information on their own that will revolutionize health care. It's the development of a single patient record that can be shared by all of a person's health care providers providing them with new context for suggesting options for treatment and prevention. It improves the cost, service, and mistake factors, but would do nothing to provide health care for Americans who cannot afford it. That is a matter of political will, not technology." —**Aaron Osterby**, State of South Australia, Department of Health

ANONYMOUS COMMENTS

The following excellent contributions to this discussion of health and the Internet are from predictors who chose to remain anonymous.

"Forget this. We have a conspiracy of all the bureaucrats who need their jobs. Hospitals need to push paper to get their cost structure up. Insurance, likewise. We are in a deadly embrace."

> "It may be wishful thinking, but I am hopeful this will be the biggest area of Internet advance."
>
> —Anonymous respondent

"The next 10 years will be marked by a series of disasters regarding the uses of ill-thought-through technologies (e-health, telemedicine), and insurance costs will be so high that government intervention will be needed. It will be common to have been injured by a failed IT-in-medicine system. There will be a strong countermovement to humanize medicine."

"Sadly, this field will continue to be a laggard."

"Medical care costs are driven by the need to fund improvements in technology and these will continue to grow at 5% to 10% in excess of inflation. As a result, IT improvements may improve the standard of care and reduce mistakes, but they are unlikely to tame the rise in costs. Also, rapidly rising costs are likely to leave excellent medical care the exclusive right of the wealthy."

"Medicine is the last big entity that has not completely adopted IT practices. Care will become more 'virtual'...and this will increase its availability, lower its cost, and if all goes well, limit hospitals/clinics to only times when they are really needed."

> "There will be a downside in more litigation, as ordinary people gain access to more knowledge about medical conditions, which makes them educated, but insufficiently expert. A little knowledge is a dangerous thing."
>
> —Anonymous respondent

"No, rising costs of health care are not going to be solved by more information; we might even see more rapidly rising costs as patients demand more and more expensive interventions they have encountered online."

"Good records management will be the No. 1 improvement, though the Internet and online is only a small part of that."

"The health care industry has been traditionally and dishearteningly slow to invest the capital in technology development. Four percent vs. over 10% in retail environment."

"Certainly, customer service in health care will change. Costs will not decrease. Malpractice concerns need tort reform, not online resources. Perception of access might change; actual access is a political issue."

"Disparities in access will probably increase without significant public investment, which is unlikely."

"Consumer preferences for health care will likely cause the share of GDP and the federal budget accounted for by health care to increase."

"Strong forces in medical and pharmaceutical and insurance sectors may affect this process more than Internet's capabilities. Complete, or at least massive, overhaul of health care infrastructure, including and reflecting Internet role, is necessary for this prediction to materialize."

"If better health information follows to patients, I can see these improvements being made. But will HMOs or other institutions control the information flow?"

"Health care may improve, but moving services online will certainly not reduce health care costs or poor customer service, mistakes or malpractice claims."

"The Internet will hardly be a silver bullet for personal-injury lawyers and the lawsuits—nor will simply overlaying a technology onto a legacy culture of poor service yield a sudden surge in warmth, compassion, and customer service. Technology is not a panacea for a broken culture."

"Efficiency is another benefit of the Internet, as is sharing of best-practices and record-keeping, comparative analysis, and second opinions."

"Only with government intervention and encouragement. The profit motive is too seductive for it to happen on its own."

"Communication is a tiny proportion of the cost of medical treatment. How can reducing this tiny fraction of the cost (and that's, in essence, what the Internet is about) alter the big picture?"

"The Internet will make a large difference in the way that people access medical care. They will be much more likely to look up information on conditions and alternative care techniques. But it is hard to see how this will reduce cost, improve service, or lead to significant reduction in malpractice."

"I am not confident that the Internet will make a meaningful difference in the overall health care system—the problems there are deep-rooted and structural. However, there will be many individual success stories of the benefits of online health."

"Bad data will still yield bad results!"

"While the medical community has successfully resisted a major IT overhaul, I see it as inevitable, given the rising costs and changing demographics in the U.S."

"This is possible, but not likely in all areas due to many legal problems, technology access problems, identity and privacy issues, regulations, etc."

"While the Internet will continue to be a source of medical information, thus empowering patients more than ever, it can't do much for the U.S. health care system as a whole."

"I think the major problems here are social, not technical. We can already do most of the things suggested here with current capabilities, but there is no evidence that the large-scale social changes that would be necessary to enable these improvements are going to be addressed anytime soon. The problems seem to be getting worse, not better."

"Other health problems may arise from the self-care, self-serve consumerist behaviors. Overall, I agree with the prediction; although, several sensational horror stories will emerge about poor self-service medical care choices."

"Needs fundamentally new processes that harness the Internet; Internet alone is not enough."

"Technology itself is never the answer. Technology may enable these changes, if there is a will to do so."

"The transition will be a very long and bumpy road. I personally am experiencing one of the bumps. There is no mechanism for removing medical records from potential use when they have been proven wrong. Records that claim I have multiple degenerative and fatal illnesses have been proven wrong through objective testing. Because the false records cannot legally be archived or removed from potential use in some way, they are used to deny me medical insurance and employment opportunities."

> "A utopian viewpoint. Building such systems may turn out to be enormously expensive."
>
> —Anonymous respondent

"I don't think it will help with customer service. And it may only exacerbate the access problems for the poor. But the

Health System Change

readily available information should certainly marginally improve care."

"There will likely be increasing uses of alternatives, although it is not clear how legal these will be (e.g., buying medications from foreign sources not necessarily approved by the FDA)."

"It will be very helpful, but not for the reasons listed...The main advantage will be for peer-support and information sharing."

"Lawyer-driven, risk-avoidance medicine will continue to prevail."

"I think the problems of rising costs, poor customer service, and the high prevalence of medical mistakes have more to do with the consolidation of the medical industry and the role of pharmaceutical companies and health insurance companies. I'm not sure how the Internet will address that problem."

> "As an HR professional, this will only happen if the insurance companies (and especially BCBS) will invest money in their equipment, their programs, and their people. With dummies at the helm of the claims processing, it won't change, and customer service will be a challenging experience for consumers. Medical research will definitely be enhanced, and virtual operations will help in rural areas, provided that they have the equipment to handle the newer technologies."
>
> —Anonymous respondent

"Health care costs will continue to increas driven by increased populations, increased population densities, a belief that illness (due to stress on the body's total system) can be solved, and that health care is like car mechanics ([i.e.,] plug and replace). This is an arrogant belief."

"This is the area in which I see the greatest potential impact of the Internet."

"Services not requiring physical interaction (e.g., lab and radiological analysis, and some aspects of primary care and internal medicine) will be removed from medical centers and outsourced to large specialty contractors."

"Health care is the one area where technology increases costs rather than decreases. Until drug companies and the health care system change their focus from handling symptoms of disease to curing disease, we will not see major change to the problems within our health care system."

"Most of the problems cannot be solved by the Internet, but by reorganization of institutions and ending of absurd U.S. economic model of health care."

"The Web will continue to be a useful tool for information-gathering on health care problems that people then use in consulting with their doctors or seeking medical advice, but the evolution will stop there."

"Technology will not fix the broken structure of the medical and insurance establishment."

"The pharmaceutical, insurance, and health care industries will feel so threatened that they will put up even more walls blocking access and use tactics that question electronic veracity. There will not be coordination; there will be no way to make corrections (see the Transportation Saftey Administration's watch list, for instance)."

"Medical information will increase, but it will not eliminate human error, underfunding, and inadequate resources."

"Technology is not a magic bullet for the problems facing health care. In fact, it has actually led to increases in costs and the extension of life beyond that which is truly humane. A true solution to the health care crisis is far more messy and more human than anyone is willing to admit. It will require concessions by attorneys, doctors, and pharmaceutical companies—concessions that I doubt will be fully worked out by 2014."

"Both the rising health care costs and its result, the lack of access to medical care, are the results of health becoming a source of profit for investors who already have money. Nothing inherent in the Web fixes that. Ditto for poor customer service. There is a potential gain in the widening of access to medical specialties for those who have the money—digital records, easy transmittal of test results, and MRIs for second opinions or consultations, etc.—but the increasing corporate seizure of what had traditionally been a private, in-person matter between physician and patient also brings with it denial of benefits, and thus of services, that act as a counterweight to those benefits for far too many Americans and world citizens."

"Legal Liability is the inhibitor to freedom of information in health care. It always has been. Read *The Great White Lie* by Bogdanich. It is so bad, people are not allowed the information to understand their illnesses much less to manage their health. The dividing line between doctors and patients is also liability. Doctors can't give information because of liability. BIG PHARMA cannot because it is marketing. We need to deliver information and choices inside of the point of care. People need tools to record their data, data visualization so they can understand trends and the implications of their choices. Disease State Management is a bureaucratic joke. If we are ever to empower people in their own care, we must bridge the learning/liability/information/telemetry gap in medicine. We should be empowered with the information to lower our health care costs by complying with care instructions and reducing our risk of health problems. It's simply a question of incentives for CORP, BIG PHARMA, Insurance, DOC, NURSE and PATIENT/consumer."

> "This is contrary to the trend of other predictions, which suggest the Internet increases diversity of views and the selectiveness of seeking information. Medical quackery is alive and well on the Internet. The health care system has critical flaws unrelated to the information highway."
>
> —Anonymous respondent

"Dead wrong. The health care industry has been remarkably slow in adopting digital technologies already—and benefit from their resistance by making it more difficult for patients to understand and question billing, etc. The industry will continue to drag its feet for fear of giving up its lucrative control. Further, medical care is by its very nature hands-on, and there can only be a limited impact by the Internet."

"People have the tools to improve their health already. Most choose McDonald's over the treadmill. The Internet will only give them access to data they will ignore."

ENDNOTES

1. See "Internet Health Resources," published by the Pew Internet & American Life Project, July 16, 2004, available at http://www.pewInternet.org/PPF/r/95/report_display.asp.
2. From http://en.wikipedia.org/wiki/HIPAA, November 22, 2007: "Health Insurance Portability and Accountability Act of 1996 (HIPAA) is a set of rules to be followed by health plans, doctors, hospitals and other health care providers. One key provision requires health plans and providers to use standard formats for electronic data interchange, such as electronic claims submission."

Part 14

Personal Entertainment

Prediction: *By 2014, all media, including audio, video, print, and voice, will stream in and out of the home or office via the Internet. Computers that coordinate and control video games, audio, and video will become the centerpiece of the living room and will link to networked devices around the household, replacing the television's central place in the home.*

Experts' Reactions	
Agree	53%
Disagree	18%
Challenge	10%
Did not respond	19%

Note. Since results are based on a nonrandom sample, a margin of error cannot be computed.

Many experts believe that a media convergence is imminent, but most of those who agreed with the premise of the prediction added caveats and elaborations. A number of experts protested that all these media toys will be available, but only to an elite group.

One expert represented the most optimistic observers, writing, "The Internet is becoming 'data electricity' and increasingly is the conduit by which information and entertainment enters the home and is enjoyed and shared with others inside and outside the home. This will probably happen before '14."

Gary Bachula, a technology development leader, most recently at Internet2, also thought 10 years was too long a timeline, unless the "digital rights management cops" put an end to the fun. He wrote, "Eventually, everything digitized in the world (movies, music, books, newspapers, etc.) will be available from the network through peer-to-peer like networks. The Net will become a giant TiVo, and will have every song, every movie, every TV show (from some point on), every sports game, every news broadcast, ever created. People will obtain it over the Net and send it within their homes wirelessly to devices that are hybrids of what we call computers and televisions today."

James Brancheau, a vice president at GartnerG2 analyzing the media industry, wrote, "Media access won't be exclusively through the Internet; it will include many types of IP networks including private (e.g., cable, satellite) and public networks (e.g., datacasting, fixed wireless). TV will lose time share to the media PC and media appliances, but it will remain central to mainstream households. The tipping point away from TV will be further down the road, perhaps closer to 2020 when our 24-year-olds turn 40." Another expert echoed these ideas, writing, "I agree with the streaming and beaming of all media, but I do not agree that it will center on the new 'hearth' of the home. I believe media will be small, personalized, and wearable. We might connect to a display system periodically, but it is more likely to be impromptu small gatherings—decentralized use throughout the home."

> "Functionality will be there, but full adoption will not."
>
> —Anonymous respondent

One detractor wrote, "Totally inconsistent with prior experience. Changes will be substantial but they will not affect all homes and new media or delivery ways will not replace old ones."

Tobey Dichter, founder of Generations on Line, also challenged the prediction, writing, "As with some of the other predictions, this presupposes an affluence that is not reality. Such broad-based connectivity requires costly subscriptions, tech know-how, housing flexibility, and an interest in leisure activities beyond television, which are beyond the capacity for a vast number of Americans." And B. Keith Fulton, vice president for strategic alliances at Verizon Communications, pointed out, "After a hard day's work, most of us want to sit and watch or listen to our devices, not interact with them."

ADDITIONAL RESPONSES

Many other survey respondents shared incisive, overarching remarks to the question about the Internet and personal entertainment. Among them:

> "They will all be available, but it's not clear that the Internet will be the delivery vehicle of choice in 2014. It is inherently less efficient than multicast media, and data volumes are still an issue." —**J. Scott Marcus**, Federal Communications Commission

> "The right model for delivering media remains up in the air. Many analysts thought videoconferencing was going to be huge—it did not turn out that way. We are still waiting for the promise to turn into a market. It could be the Internet, but frankly it was never set up to deliver this type of content. There are companies out there right now trying to figure out which model will work. I am not yet convinced streaming is the answer." —**Bradford C. Brown**, National Center for Technology and Law

> "Centerpiece of the living room? Puhleeze. Still, the television is not now the central place in the home." —**Moira K. Gunn**, host, *Tech Nation*

"I will add that the TV will still play a central role as a display." —**Mike Kelly**, America Online

"There is growing convergence of electronic media that will make the computer the primary entertainment device in homes, replacing the many different devices currently in use." —**Gary Kreps**, George Mason University, National Cancer Institute

> "The Internet will play a big role, but there will always be a place for traditional print media (or at very least, books, especially fiction). Even George Jetson used paper from time to time."
>
> —Brian Reich,
> Internet strategist
> for Mindshare Interactive
> and editor of the political blog
> campaignwebreview.com

"All this will be available by 2014, but will not be as universal as, say, DVD players are today. Nor will it be the only form of connectivity. But it will be an option for many, and the devices that can network around the house will be mainstream." —**Benjamin M. Compaine**, a communications policy expert and consultant for the MIT Program on Internet and Telecoms Convergence

"This is at the heart of my new book. I think an attempt to identify the black box through which media convergence occurs is misguided. We live in a culture defined as much by divergence of communications technologies as we do one defined by convergence. Culturally, we already live in a world where messages, ideas, brands, characters, stories, and content flows fluidly across media platforms and people make connections between information gleamed from many different media sources. We will, of course, see various attempts to technologically integrate those flows. Our cell phones are more and more the digital version of the Swiss Army Knife, but I don't think this will ever be reduced to a single channel of communication in the way you describe." —**Henry Jenkins**, MIT Comparative Media Studies, author, *Convergence Culture*

"This is already happening to some extent with the rise of broadband. But TVs might well become more like computers, rather than computers becoming more TV-like. Perhaps the TV will become part of that network hub for media. The problem is the high cost of these systems, which should come down to some extent by 2014." —**Mark Glaser**, Online Journalism Review, Online Publishers Association

"Without some radical changes in wireless and the amount of bandwidth available, there will be a need for radio, TV, and other electronic mass media. The present wired and wireless Internet basically is lower-bandwidth than telephony, without building an 'electronic superhighway,' which all politicians love but no one wants to pay for." —**Mike Botein**, Media Center, New York Law School

"Excluding, of course, the homes of the poor, who will continue to get information and media via television and radio. The digital divide will deepen not only between the wealthy and the poor, the wired and the unwired, but also the educated and the uneducated." —**Mack Reed**, Digital Government Research Center, USC

"Such systems will be possible, but the question is whether people will use them. People are often slow to change things in the home." —**David Tewksbury**, University of Illinois at Urbana-Champaign

"Most homes (like most well-run businesses today) will use a variety of feeds and deploy a variety of networks to avoid critical reliance on any one." —**Philip Virgo**, secretary general, EURIM

> "Someday, all telecommunications to and from the home will be digital and packet-switched, but whether it will be in 10 years or 50 I can't say. And people will still be couch-potatoing in front of moving images on a video screen in the future, digital or no."
>
> —Tom Streeter, University of Vermont

"This will certainly be the case for those who can afford it. One of the opposing pressures will be the rapidity with which each technology makes the previous 'solution' obsolete, and consumers' eventual resistance to the cost of 'keeping up to date.' It used to be you'd buy a camera and use it for 1 or 2 decades; now, the digital camera you bought 2 years ago is already screaming to be replaced. The same goes for MP3 players, DVD standards (and future storage options), screen technology, and so on." —**Rose Vines**, freelance technology journalist, *Australian PC User, Sydney Morning Herald*

"All media will stream through the Internet, but alternative sources will still be in use by a generation weaned on them. Devices that control household data-flows and link devices will become more ubiquitous; however, where in the home they will be and what component they are attached to is still unclear." —**Jonathan Peizer**, CTO, Open Society Institute

"Yes…*but*. You're talking about a controller, not just a receiver/display which was TV set's role for 50 years. The current cable/satellite STBs fulfill part of this role, although they are still primarily receivers. The future media-center computer, which will be widespread but not universal by 2014, will be a steppingstone toward the on-demand environment you're describing." —**Gary Arlen**, Arlen Communications Inc.

> "Part of this prediction is true—that all media will flow through the Internet in some fashion instead of being through proprietary channels and networks. However, the television is already not central in many households, as many households have multiple televisions throughout. Enhancements to traditional TVs and audio systems will simply take the place of or the place near existing devices, just as the DVD is replacing the VCR but is only marginally increasing where movies are watched."
>
> —Dan Ness, MetaFacts

"Whatever you call it, a big HD screen with good audio around it will still be the central place of home and you can watch channels on it and check/display other information resources. So why not call it a 'television'? And the entertainment systems (and other household appliances) will be networked, together and to the outside. But I don't think it will be the Internet. It will be local and community IP-based networks, private networks with and to friends. There will be a link to the Internet somewhere, but as said, I don't believe it will be the prime source of the data coming in." —**Egon Verharen**, innovation manager, SURFnet (Dutch national education and research network)

"There are too many easy, profitable, and reasonable ways to write personal computers completely out of the individual media delivery channel that I predict they will be written out in all but hard-core-geek homes. There is no easy or profitable way, however, to write server computers out of the distribution channel. The Internet may come in over the electric wiring in the house, in addition to twisted pairs or cable. Dial-tone-level control panels mounted on the wall or on the media delivery device (that are as easy to use as a microwave oven control panel) will deliver subscribed-to access to whatever entertainment the family wants. Or, yes, I'm going to say this: another 'clicker' to lose in the couch." —**Elle Tracy**, The Results Group

"I agree this will happen. I think it'll take till the 2018–2020 time frame, though. Look how long it took HiDef to replace NTSC. A huge installed base is a powerful hysteresis generator." —**Mike O'Brien**, The Aerospace Corporation

"The only thing I would add is that the notion of what is 'Internet' will become irrelevant. Meaning that the distribution mechanism and devices will be largely irrelevant. What will matter is that media will be interactive, largely pervasive, and subject to increasing user control." —**Sam Punnett**, FAD Research

"The computer will not be the 'centerpiece of the living room' or replace 'the television's central place in the home.' The computer will be located in the home office or basement, and the integration will be wireless and invisible. The television will be replaced by screens that can stream a variety of media or applications, and will be located in several rooms of the house." —**William Stewart**, LivingInternet.com

ANONYMOUS COMMENTS

The following contributions to this discussion about the Internet and personal entertainment are from predictors who chose to remain anonymous.

"I agree that we'll have this potential and some segment of the population will see this environment. But the majority of the public will not have the resources or level of involvement to consume media in this fashion. The newspaper will continue to exist, as will print magazines and portable media devices."

"While I agree that media will move online, I do not believe that PCs with standard operating systems will migrate to the living room. Instead, I think smarter appliances will assume that role, since they are much less expensive, and easier to use than PCs are. These appliances (e.g., TiVo on steroids) are more likely to replace television."

"Too many 'ifs' for a 10-year outcome."

"I agree, but wonder if the common distribution network will still be called the Internet and if the screen that people watch will be called a computer or a TV. These are just names—the experience will be similar although the viewer will have more control over what he/she watches, when and where."

"This will be an option available to some, and it will be part of a range of media services that people can choose. This range of options and media will be incredibly rich, diverse, and difficult to characterize."

> "Hey, I am working on it!"
>
> —Anonymous respondent from a company that makes software and hardware

"While computers will be the hubs of the networks, I prefer to take the view that the hub will control the various devices. Many have often held the belief that all data will converge into one device and that one device will take the place of all other devices. Your question itself is contradictory because, on one hand, you contend that the computer will be the centerpiece; on the other hand, you also contend that there will be various networked devices around the home. Can't have it both ways. Yes, computers that are networked in the house will function as a system. No, they won't take the place of the television, although they will likely be controlled by networked computers. And no, we won't all gather around the computer in the living room. The notion that 'a' computer will be in the home is the first flawed assertion in the statement. Summing it up: The computer will converge the data so it can then be distributed to myriad divergent, smart, computer-like devices with various uses around the home."

"I'm not entirely sure it will be the Internet, but I do expect convergence of media to a packet-based infrastructure, and the emergence of home networks that integrate and transform the current generations of computers, entertainment devices, and appliances."

"Technology is there, but the cost of running 'fiber' to the home will not change slow movement in that direction."

"Largely true. The exception may be the local newspaper. The Internet hasn't yet figured out how to do this well."

> "The predictions are all pretty much correct, but the time frame is overly optimistic. These things will happen, but probably not over a 20-year time frame. And whether TV, as a passive entertainment experience, gets replaced or simply becomes on element in the overall experience of multimedia, remains to be seen."
>
> —Anonymous respondent

"Cable and satellite will still be distributing a lot of the video to the home in 10 years, though it will all be digital."

"The entire home entertainment system won't change that quickly. These things are more expensive for the average user to upgrade than some may realize."

"Yes, some form of networked device will replace TV; no, there will be other sources of media as well."

"The Internet will be one of many pipes into the home."

"Families will react and challenge the pervasiveness of the 'centrepiece in the living room.'"

"Except, of course, that the big screen will still have pride of place."

"The need for this kind of networked media is low so that television will still be the central entertainment system. Personal devices aimed at individual use will be streaming."

"The convergence of computers and television as a technology will occur, and that 'entertainment center' will be the new TV in 'the central place.'"

"But just as television did not replace radio and print media, this entertainment center will not replace all other forms. Likewise, as TV has spread throughout the home, it is wrong to place the entertainment center as 'the center of the living room.' The wired house will operate on many levels throughout the home."

"No more than 50% of the population will experience this due to cost, and the slow roll-out of "big broadband" to the home."

"Agree, except that you won't think of it as the Internet. It will probably be packetized, but you'll have certain streams prioritized to avoid interruption…and third-party companies will have to pay to be prioritized in the bitstream."

"Television will be replaced, but I'm not confident that computers will reign supreme. It may be cell phones or some other device, just being invented."

"There will be a large and very noticeable increase in the amount of media accessed online, streamed, and downloaded for on demand use. But in the average American household, the TV will still be the centerpiece."

> "Are mobile phones 'the Internet'? Why 'the living room' and why the 'centerpiece'? What of ubiquitous computing?"
>
> —Anonymous respondent

"Not by 2014, but one day as the technology can do it. America has a lower broadband penetration than developed Europe, so if this does happen, it will happen in Europe first. However, there are massive regulatory hurdles in the way."

"Not via the Internet—too expensive—but via IP-based services mostly on closed systems from the network provider. I agree with the home-networking point."

"For a small percentage of the world's population. The rest will probably be even less connected as the wealth continues to concentrate in the hands of the few who can afford these luxuries."

"Yup, I expect nearly everything we do for communication will be online soon. Might be later than 2014 before everyone has retooled, but I believe it's coming. Will it replace the TV? No, the TV itself will be networked and have far better display capabilities."

> "As long as the visual senses can be stimulated, the Internet will enjoy popularity. Whether it be TV or PC, a visual display screen has to be the focal point for personal entertainment."
>
> —Anonymous respondent

"The promise of broadband will remain elusive and will not deliver; the television will never be replaced by the computer."

"I only challenge the timing. The cost at the household level for this sweeping change may be too high for widespread penetration within a decade. If costs are minimal and usage is simple (an important factor), it is likely to take hold."

"I don't think the TV will disappear anymore than the radio disappeared with the advent of TV."

"There will be choices of how to receive and manage this content; the devices will be networked."

"That's certainly what my company hopes!"

"It will happen by 2009."

"'All media'? All? No. Some/a lot? Yes."

"Yup, and I can't wait. Some of this is already possible today—but is expensive—so beyond the reach of most."

"Newsprint isn't that easy to kill."

"It's probably true, even though it doesn't sound like a pleasant prospect."

"[There are far] too many assumptions wrapped up in this question. The television monitor, as the largest display in the home, and the one around which people can gather for a social experience, will not be displaced by these other devices. The home server you describe will be the functional centerpiece of the home network, but it does not have to be the centerpiece of the living room and will be most successful when it is invisible and its functions transparent. It may even be distributed across several networked devices, as we see PVR (personal video recorder) technology already becoming common in multiple boxes, notably the cable/satellite controller and DVD player/writers. Television broadcasting, as the sole or main means of getting content onto the TV display, will share access to it (and to all the devices) with broadband and other physical media such as portable memory (like tiny flash drives for music players and cameras), DVDs, and massive in-home hard drives. But broadcasting will remain an efficient and effective means of mass communication, just as physical delivery of paper mail continues in the age of TV, radio, and e-mail. It changes, but it doesn't die."

> "Already happening. Broadband penetration is over 50%, the power grid adaptation to a broadband network will accelerate this; devices will become simpler. This one's a no-brainer."
>
> —Anonymous respondent

"I agree that digital media will be transported by the Internet and will have taken the place of the TV; I think textual media will still persist in a combination of digital media and paper."

"Consumers will continue to demand high fidelity in sound, pictures, and other attributes of their information and entertainment media. Some of this is best distributed by dedicated networks (like broadcast or satellite radio and TV); some can be downloaded over the Internet for 'playback' through high-fidelity devices."

"Other communications are more efficient. Try out XM radio. If I have XM radio, what do I need to do that over the Internet for? Broadcasting services are not well suited for Internet design. But then there is that SUN vision of the future where every move ever made is downloaded to a box in your home, which the individual views on demand. I think people still like entertainment services—iPod does not replace radio completely because people like to hear new things; they like to hear what they have not heard before; they are bored by their own playlist."

"A television is just a display. I don't see displays going away. STBs have had chips in them for years and no one seems to have noticed. So I don't see a whole lot of change—TV as a display will remain a central part of the home."

"TV is already dead. Kids are spending more time on games in network game worlds. The largest demographic of gamers is women over 40. Legacy home-entertainment systems are the ball around our ankle. We are seeing a transformation of media and cyberspace. It happened 10 years ago. Didn't you see it? Are you there now? Online game worlds. Alpha World may have failed, but it was an indicator species. The media-and-cognitive landscape has shifted. Convergence is a false notion. We are seeing divergence, new systems, processes and products, not combinations of the old. McLuhan said it of course: 'Rear-view mirror.' The capital 'I' Internet is dead. Networking everything through all kinds of methods is where we are. TIVO is an indicator species too. Watch the 8-year-olds; it is for them what it is not to us. They are the answer to how it is now and how it will be in the future."

PART 15

CREATIVITY

PREDICTION: *Pervasive high-speed information networks will usher in an Age of Creativity in which people use the Internet to collaborate with others and take advantage of digital libraries to make more music, art, and literature. A large body of independently produced creative works will be freely circulated online and will command widespread attention from the public.*

Experts' Reactions	
Agree	54%
Disagree	18%
Challenge	9%
Did not respond	20%

Note. Since results are based on a nonrandom sample, a margin of error cannot be computed.

According to a December 2004 report by the Pew Internet & American Life Project, artists and musicians have embraced the Internet as a tool to improve how they make, market, and sell their creative works. They use the Internet to gain inspiration, build community with fans and fellow artists, and pursue new commercial activity.[1]

Despite the upbeat feelings of artists themselves, many experts begged to disagree about the prospect for a flowering of creativity. The idea that independently produced creative works will command widespread attention was almost universally derided as "utopian," "pie-in-the-sky," or simply "hype." One respondent wrote, "Music, art, and literature. Yeah, right. The only thing broadband will bring to the public is uncensored reality schlock shows and porn." Another expert wrote, "Humanity has had books for hundreds of years, but does not have universal literacy. Creativity may bloom, but that does not mean it will be seen or appreciated by all."

> "Predictions of a new Age of Creativity driven by the Internet are no more likely to come to pass than similar predictions made in the early years of television."
>
> —Jorge Reina Schement, director of the Institute for Information Policy at Penn State University

One respondent wrote, "The Internet overcomes the simple problem of disseminating information, but it vastly increases the problem of overcoming information clutter and overload. Marketing and publicity remain critical, whether provided by today's record labels, by a completely altruistic co-op of like-minded artists, or anything in between."

However, there were observers who believe that an "Age of Creativity" is possible. For example, one wrote, "Modern art was largely spurred by a reaction to photography. Artists adapt to new media, and will adapt to the Internet."

ADDITIONAL RESPONSES

Many other survey respondents shared their elaborations on the question about creativity and the Internet. Among them:

> "That's a nice way of putting it. I don't know that we'll have a 'creativity society,' but there will be more opportunity for bottom-up media creations." —**Douglas Rushkoff**, author; New York University Interactive Telecommunications Program

"I agree, but there is a serious risk from overprotection of intellectual property." —**Peter Levine**, deputy director, Center for Information and Research on Civic Learning and Engagement, University of Maryland

"Only if we can resolve the current differences between those who wish a return (including academic recognition) from past creativity and those who wish to build on the creativity of the past." —**Philip Virgo**, secretary general, EURIM

"I don't think the Internet helps art that much. Great art tends to be the product of individuals rather than groups, even when people are in immediate proximity. The one exception, here, is computer gaming; some of which will begin to deserve the term 'art' by 2014." —**Gordon Strause**, Judy's Book, a social-networking Web site

"The creative already create. There might be some more collaboration in certain areas, but the Internet cannot change a person's native talents and interests." —**Jonathan Band**, an attorney specializing in e-commerce and intellectual property with Morrison and Foerster LLP in Washington

"The digital medium simply allows far more creative collaboration because manipulation of bits and bytes, as well as their dissemination, is far easier." —**Jonathan Peizer**, CTO, Open Society Institute

"The more there is out there, the more important 'branded' things become. There is no way for an individual to sift through millions of songs or art files; as a result we will increasingly rely on the 'critics' as gatekeepers, or less appealingly, the record labels and TV moguls as gatekeepers. While there will undoubtedly be an increase in independent works, most of this will be 'vanity work,' and command little public attention. After

> "It happened with the printing press and the PC and is already happening online. New literacies and new means of distribution mean new outlets for the boundless human urge to create and share creations. Business models are another question."
>
> —Howard Rheingold, Internet sociologist, writer, spreaker

all, what good is watching television if you can't talk about it at the watercooler the next day. Part of the purpose of arts and entertainment is to provide a common cultural bond."
—**Vikram Mangalmurti**, Carnegie Mellon University

"I do think this will happen. Applications like Garage Band, when networked, will make artistic collaboration as commonplace as literary collaboration became with the advent of e-mail." —**Kevin Featherly**, news editor, *Healthcare Informatics*, McGraw-Hill Healthcare Publishing

"I have great faith in the creative spirit, but art is ultimately a product in our society, and what we have is a dis-integrating marketplace (one that actually militates against integration). The proliferation of media is also a fragmentation of taste and interest, one we already see in the music market, the TV audience, and so on." —**George Otte**, technology expert

"More and more searchable pools of amateur music and art and writing will emerge. But it will still be hard to get attention. The desire for a shared, numbing, pulsing branded experience will continue to overwhelm."
—**Susan Crawford**, a policy analyst for the Center for Democracy & Technology and a Fellow with the Yale Law School Information Society Project

> "I do not believe that a large public will attend to the new art. Rather, the Internet will function like a big city in which avant-garde and strange arts and events can do well because the large population produces the critical mass necessary to sustain them."
>
> —Stanley Chodorow, University of California–San Diego, Council on Library and Information Resources

"Don't expect human behavior to change fundamentally. More people will take advantage of digital libraries and online cultural resources, but the increase will not be dramatic." —**Terry Pittman**, America Online, broadband division

"Mainstream publishers will decline in importance." —**David M. Scott**, Freshspot Marketing, *EContent* magazine

"We are used to seeing tinkerers and amateurs playing an important role at the outset of new technologies, but the Inter-

net has enabled this kind of amateurism at a much broader and later stage of development. Along with new forms of work and social organization that favor individual creativity, I think we are on the cusp of a new creative renaissance."
—**Alex Halavais**, State University of New York at Buffalo

"The Internet lowers the threshold for participation in these activities and unlocks creativity that people never knew existed, or had access to." —**Brian Reich**, Internet strategist for Mindshare Interactive and editor of the political blog campaignwebreview.com

"This is already happening, with greater opportunities for collaboration and sharing of information." —**Gary Kreps**, George Mason University, National Cancer Institute

"I could agree to this prediction with caveats or disagree with reservations. The tools will make this possible and many will take advantage of it, just as they already are with digital photos, making collages, etc., and sending them around to family and friends. But very little will get the key (widespread) attention from the public. Creativity at that level is not a common attribute. Materials that will attract a mass audience will still be quite finite, due to time and cost.

> "Creative people will be creative on this medium; some others may find it who would not have—others will just use it the way they are using it now."
>
> —Cynthia Samuels,
> Center for American Progress

These networks will remove some institutional barriers, but it will be analogous to authors today who self-publish—a few get attention and 'break out.' But most stay obscure not due to inability to publish but because they are not of mass market (widespread) caliber." —**Benjamin M. Compaine**, a communications policy expert and consultant for the MIT Program on Internet and Telecoms Convergence

"I agree with this as long as we keep our expectations in check with respect to how many people will be creating inventive works and how many will be paying attention. There is little evidence that the public at large, en masse, is likely to be attracted to 'creative works.' Look at 'reality TV.' There will be strong subcultures of gifted artists and probably larger audiences for the work, but the public in general is not likely to be much more enlightened than it is now." —**Gary Chapman**, LBJ School of Public Affairs, University of Texas at Austin

"Maybe. In the end, the creative process still requires money (funding) to cycle through it. It remains to be seen whether or not artists would be willing to contribute their independent works to a collaborative commons without a mechanism to be paid. Moreover, the copyright law is still relevant and would need to be addressed in some way." —**Bradford C. Brown**, National Center for Technology and Law

"Not in 10 years. I also note that the sciences are not included in the prediction. Creativity is not solely confined to people that call themselves artists." —**Robert Lunn**, FocalPoint Analytics, senior researcher, 2004 USC Digital Future Project

"I don't think that the Internet will prompt more individuals to be creative. Rather, it provides a new medium and outlet for those already bent on creative expression." —**Michelle Manafy**, editor, Information Today Inc., *EContent* magazine, *Intranets* newsletter

> "I think the 'Age of Creativity' is already upon us, in the sense of a flood of a huge variety of material of hugely varying quality. Things will still be like that 10 years from now. What and where the really good and/or popular stuff is coming from will depend on other factors besides technology."
>
> —Tom Streeter, University of Vermont

"Independent works will be freely circulated, but I don't think they will command widespread attention from the public." —Joo-Young Jung, University of Tokyo

"Certainly, my own work celebrates the explosion of grassroots creativity we are seeing as a result of new media tools, new channels of distribution, new modes of expression, and new communities for collaboration. Yet, I can't overlook the serious legal battles over copyright that are taking place, which seem to be working desperately to contain grassroots creativity. People are fighting for their right to participate in their culture and to make use of the core materials of their heritage. The outcome of those struggles will define the kind of culture we live in much more than technological changes will." —**Henry Jenkins**, MIT Comparative Media Studies, author, *Convergence Culture*

"Collaboration will be widespread but not universal. The biggest challenge is the business model: creativity for its own sake or for money? Social values will affect this process. Also, production values. Will viewers want to see homemade programs that look or sound less slick? Yes, some will want such authenticity." —**Gary Arlen**, Arlen Communications Inc.

"It's all true, except to say that the 'body of independently produced creative works' will command widespread attention, but only a small proportion of the individual works within that body will break through to the masses of people the way that the most popular of mass media regularly hits tens of millions of people today." —**Peter Eckart**, director of management information systems, Hull House Association

> "Unfortunately, a lot of work that is termed 'creative' really isn't. The amount of good creative work produced by human beings has held relatively even output over the course of history. It's about people's brains and spirits, not about the tools."
>
> —Oren Schlieman, InfoGrafik Inc.

"Most people are just busy living their lives and won't have the time to either produce or consume this 'creativity.'" —**Rob Atkinson**, Progressive Policy Institute (previously project director at the Congressional Office of Technology Assessment)

"Unless you solve the fair-use aspects of online copyright, this prediction will not happen. Today, Big Media owns the debate, the lawmakers, and the technology companies. Until someone champions fair use online with equal ferocity so that solutions can be worked out to benefit all parties, there will not be enough material online to make your prediction happen. The other issue is cost. If Big Media wins the argument, even the incremental cost to buy access to materials needed to foster creativity will be a barrier to your prediction. Money will limit things, not the creativity of the creators." —**Tim Slavin**, ReachCustomersOnline.com

"Technology doesn't automatically create creative works, although it will enable more individuals to create. The largest portion will be experimental and considered substandard, much as today's blogs and personal Web sites are today. Still, there will be an incremental increase." —**Dan Ness**, MetaFacts

"This is true today. In literature, the blog has recreated the old French salon. But, more intriguing to me, representational art is also benefiting. Some years ago I made a strong and heartfelt prediction that online, 3D virtual-reality, shared spaces would be slow to be adopted because almost no one in our society grows up with the sort of artistic training needed to build attractive spaces. Then Second Life came along, and brother, was I wrong. Even though public art education is so bad that it makes those who complain about education in the sciences look like a bunch of whiners, the stuff built by 'the commons' in Second Life is a complete knockout. *Snow Crash*, here we come." —**Mike O'Brien**, The Aerospace Corporation

"Look at all the creative, high-impact, citizen produced stuff that (came) out of this 2004 Presidential Election. It was enlightening, funny, and maddening. Everything creative work should be." —**Leonard Witt**, PJNet.org

"Creativity is a function of inspiration and ability. Most folk will not become more creative because they have faster networks. They will become more informed though. And may virtually visit places where creative outputs are stored."
—**B. Keith Fulton**, vice president, strategic alliances, Verizon Communications

"The loss of the intellectual commons has been sped by the growth of the Net, though not driven by it. The increasing legal standing of corporations as persons has been far more significant. However, I think it is more likely that these corporate forces will figure out how to inhibit and profit from the flow of information than that it will become freer. The last line of the prediction is especially problematic. Interactivity is the essential ingredient for a creative product to command widespread attention. Corporations are likely to continue to be the ones with the capacity to bombard the citizenship with repeated exposure to a cultural product." —**Alec MacLeod**, California Institute of Integral Studies

"Oh well, everyone will be creative and nobody will read it or watch it. If freely available work gets the kind of distribution that challenges media company profits, then it will probably be crushed in some way. How do the freely available people raise the money for their projects, how do they get the projects out from the morass of other projects?" —**Jon Marshall**, University of Technology, Sydney

ANONYMOUS COMMENTS

The following contributions to this discussion about creativity and the Internet are from predictors who chose to remain anonymous.

"I think people will have more access to other works, but I question whether that necessarily leads to greater creativity (or merely more highly derivative and not very imaginative work)."

"The playground will be open. This is one of the most exciting aspects of the network space."

"An 'Age of Creativity' is too much to hope for."

"I think that most people will still consume media in a largely passive way (or interactive only in form of playing videogames)."

> "Much as the word processor made it possible for anyone to make crap look good, so the Internet makes it easy to distribute crap. Widespread attention will continue to be given to quality, and the Internet will not create talent. If anything, it will make the good stuff harder to find in the sea of crap."
>
> —Anonymous respondent

"As governments cut back on expenditures, libraries will be defunded. Media conglomerates will extend copyrights to preclude their property moving into the public domain. The public will spend more and more of its time consuming media experiences, rather than creating their own."

"I agree with much of the statement until the last phrase. Widespread attention is less likely because there will be more works in circulation."

"Creativity does not depend upon the Internet. It depends upon creative people."

"Creativity won't be changed by more people and groups spending more time together."

"It all depends on the IP laws and how they change (or not)."

"We simply cannot predict trends in creativity. I think new forms of creativity will emerge, but will there be 'more' or 'less' of it or will it be 'better' than what's been done in the past? I think there will be more opportunities for marginalized artists (and nonprofessionals) to promote their work, but that doesn't mean anyone will want it."

"Zipf's law indicates that only a small percentage of material created will be of widespread interest. All societies have more readers than writers."

"These products, while likely, will not compete with corporate-generated content. That is a shame, but I think it is true."

"The prediction is part correct. The creativity will flourish and there will be collaborative creative products that will circulate widely, but they will not command widespread attention of the public, which will still focus on the products of large studios/conglomerates who will dominate through advertising and marketing power."

> "The barrier will be copyright issues. Much creativity comes by building on prior work. The current trend in technology and law is to make that increasingly difficult. It see that as seriously impending this vision."
>
> —Anonymous respondent

"It won't make people more creative, or give them more time to be creative."

"There will be several high-profile activities like this, and lots of people who find ways to take what they find around the Internet and bring it to their worlds of action. Some of what they bring will be fallacious, hurtful, and dangerous, much will be benign, some will be truly wondrous, but all on a local scale."

"The library and the museum become worldwide as access to these continues to go online. I do wonder, though, who is going to pay for the museum itself."

> "I wish. You wish. But people tend to collaborate on more banal things—games, listening to music. The creative process is still often very private. Maybe by 2024."
>
> —Anonymous respondent

"I agree with everything except the digital libraries part; because of increased lockdown of copyrighted images, it is very dangerous to assume that the creativity, though possible, will be legal. The Gray album proved to us that technology can not shut down distribution of this type of creativity, nor can it ensure that creators are compensated for their labor. It did ensure that the works created and desired by the public to be created were illegal."

"The technology will be there, and a few people will use it in the way described; but most will remain relatively passive consumers."

"Lots will get created, but most will be ignored. Production values (as in today's games and movies) will continue to escalate so as to keep most such creations in the realm of 'folk art.'"

"I don't think we have any evidence of this right now, and fears about copyright issues will stifle innovation and creativity."

"Sorry, this just seems too pie-in-the-sky. Some people will utilize the creative potential of the Internet in these ways, but the general public is unlikely to be elevated to a more creative level, or to appreciate independent work."

"This is already happening." —*The response of many anonymous participants*

"The Internet contributes to a high level of conformity, shared images, shared information, and does not promote novelty or creativity in the manner suggested."

"Copyright law will squash many attempts to do this. And most Americans won't use these technologies to be creative, but to make money and sell useless crap."

"Absolutely. Already in play and more will evolve...one of the greatest legacies of the Internet. We just have to use it wisely."

"It's happening now, and will only increase as kids who have grown up with technology naturally make it part of their creative work."

"The sheer volume of data on the Internet creates too much information overload. Ushering in an Age of Creativity? No, just providing new avenues for some to do independent productions."

"This presupposes that digital libraries will be created to support art, music, and literature. That is not where the money is going for digital libraries. In fact, the money for digital libraries has largely dried up in the wake of funding for 'security.' Moreover, if 'a large body of independently produced creative works' is present in the Internet, then tools will need to be created to provide better organization for the Internet, or the public will not be able to locate such works quickly or easily. Getting 30,000 hits doesn't make it easy to find something that interests you."

"It will be free and fee. Creativity (if we can get out of our own way). Distribution (circulated online and off). Attention (increasingly fragmented—lots of small audiences). Emergence (memes will travel from the edge to the center of culture and society quicker). Public attention will be a distraction. Everything but news. Me, me, me. The commercial becomes personal. The personal becomes cultural. We keep spiraling. The Age of Creativity is required across all disciplines and especially among art, science, and culture. If not, we are doomed as a civilization. The technical must become human. Techne should be understood as the act of bringing something into the world: a human process. Creativity (e.g., adaptation and survival of the species) is dependent on the linkage of art and science and culture. Religion: not too sure what to say about that. I studied with Gilder and West and Chapman at Discovery.org. Intelligent Design?"

"More works created hardly means the same thing as more creativity."

"I sit on the board of a cable TV station and this is the focus of our current strategic planning. We envision interaction and incubation of digital artists and entrepreneurs."

"If the idea is that the age-old formula can somehow be modified so that people don't need a profit and promotional incentive to create works that achieve real distribution, then this prediction is wrong. There will always be financial struggle in producing something that requires artistic investment. The idea that improved networks will somehow lick this divide is not necessarily the case. However, it will probably make the overall market a little better for independents. Remains to be seen... How long did it take for publishing to go from Gutenberg to Random House (corporate consolidation)? That's the question you ought to consider."

> "These exist now and haven't stimulated creativity, as we know it. It is the sensory visualization of things (e.g., theatre, architecture, etc.) that still remain real and in need of human interaction. The Internet is a secondary surrogate used when you can't be there in person."
>
> —Anonymous respondent

"Already happening. The success of the movie, *Sky Captain and the World of Tomorrow*, based on a Mac-made prototype e-mailed around Hollywood...people making small films with home software and putting them on iMovie, etc."

"Creativity yields little interest from normal people, only to intellectuals. The center of the population is driven to fashionable products and sports."

"Enclaves of artists will collaborate, and their work will command attention from the small group of people who follow highbrow arts."

"While the means are available, you cannot influence a person's creativity or intent. In some ways, technology may inhibit those who are not comfortable with it."

Endnote

1. See "Artists, Musicians, and the Internet," published by the Pew Internet & American Life Project, December 5, 2004, available at http://www.pewInternet.org/PPF/r/142/report_display.asp.political lines.

PART 16

INTERNET CONNECTIONS

PREDICTION: *By 2014, 90% of all Americans will go online from home via high-speed networks that are dramatically faster than today's high-speed networks.*

Experts' Reactions	
Agree	52%
Disagree	20%
Challenge	8%
Did not respond	20%

Note. Since results are based on a nonrandom sample, a margin of error cannot be computed.

According to the February 2004 survey of the Pew Internet & American Life Project, 34% of all adult Americans have access to broadband either at home or in the workplace. Much of the growth in broadband adoption at home is attributable to users' unhappiness with the dial-up doldrums—that is, people growing frustrated with their slow dial-up connections. Price of service plays a relatively minor role in the home high-speed adoption decision.[1]

Although slightly more than half of the experts agreed with this prediction, few were specific in their responses about why they think high-speed access will roll out to most homes by 2014. Many seemed to aspire to these heights of connection, citing more hopes than facts. As one expert wrote, "Yes, but 90% isn't good enough. We must do away with the digital divide entirely if we are to become a truly advanced culture."

More typical was the expert who wrote, "Whoa. First you have to get 90% online. I don't think that is possible given current trends. I do believe broadband will be widely used, however. Still, will it be faster than today? Only if we can come up with novel ways to make that pay for itself." Another wrote, "Only if the government subsidizes the great divide between the haves and the have-nots with respect to computers, computer training, and the cost of access."

> "It will take longer to reach 90% unless this becomes a public-works project like the Interstate highways."
>
> —Anonymous respondent

Additional Credited Responses

Many other survey respondents shared incisive, overarching remarks to the high-speed of networks question. Among them:

> "Ninety percent is too high. Penetration of wired phones is only 94.2%. Adoption will probably not be significantly higher than take-up of PCs, which after all these years is still well below 90%." —J. Scott Marcus, Federal Communications Commission

> "Agree with most. Largest not-online group is elderly. But many of the elderly of 2014 today are younger and more likely online and will stay so. Approaching 90% is thus feasible. And high-speed—relative to dial-up—is also on a trend today to

get us close to that proportion. I'm more doubtful that 90% will have high-speed connections that are 'dramatically faster' than today. Many will, but not all. So two of the three parts of this prediction could happen." —**Benjamin M. Compaine**, a communications policy expert and consultant for the MIT Program on Internet and Telecoms Convergence

"Not unless someone wants to pony up about $100 billion for a national information infrastructure." —**Mike Botein**, Media Center, New York Law School

"Without substantial government assistance, this will not happen until ISPs and/or telecom companies find the cost of providing all homes with broadband is worthwhile in terms of the data they will mine." —**Lois C. Ambash**, Metaforix Inc.

"I don't know if it will be 90%, but certainly 50–75%." —**Jonathan Peizer**, CTO, Open Society Institute

"Ten years is too short, and the networks won't get much faster by then. But lots of people will have broadband at home." —**Susan Crawford**, a policy analyst for the Center for Democracy & Technology and a Fellow with the Yale Law School Information Society Project

"Agree—will be a mix of wired and wireless. Broadband connectivity will be ubiquitous." —**Terry Pittman**, America Online, broadband division

"Most access to the Internet will be via wireless networks, especially as cities begin to establish Wi-Fi grids. Just as many Americans are foregoing their landlines at home for cell phones, I suspect that they will give up their broadband for wireless." —**Alex Halavais**, State University of New York at Buffalo

> "We're almost there. Ironically, it's our competitive marketplace that holds us up. While we have networks that fail to integrate and be interoperable, nationalized approaches in places no more tech capable (e.g., South Korea and Italy) are really further along in this respect."
>
> —George Otte, technology expert

"It's too extreme a prediction. A lot more people are going to be on home access broadband, and it's very likely that this will involve wireless access or access through power lines. I just don't think it's possible to have 90% of ALL Americans on these new types of high-speed networks in 10 years. Perhaps 90% of current home Internet users, but not 90% of total Americans, from their homes." —**Robert Lunn**, FocalPoint Analytics, senior researcher, 2004 USC Digital Future Project

"No. We're not going to get there. This will be true in South Korea, Japan, and perhaps Singapore, but not in the United States. Recent public policy has already doomed this dream to failure in my view." —**Kevin Featherly**, news editor, *Healthcare Informatics*, McGraw-Hill Healthcare Publishing

"Half-Meg broadband penetration in the U.K. stands at over 95% and new homes are starting to be built with 100Mbps, CAT5 enabled. The USA will make a national commitment to FTTH [fiber to the home] within 5 years." —**Steve Coppins**, broadband manager, South East England Development Agency, Siemens

> "Due to regulatory and monopolistic limitations presented by the FCC, RBOCS, and cable industry, the U.S. will fall further behind many industrial countries as far as overall usage and Internet speed. There will be some overall improvements in mobile computing primarily produced by wireless technologies."
>
> —Pat Murphy,
> CAPE.com Internet services

"Not sure it will reach 90%, based on historic growth patterns. The wild card will be projects such as CENIC, which offer government or government/corporate collaboration to build and maintain and train users for such high-speed access." —**Gary Arlen**, Arlen Communications Inc.

"Absolutely. We will all have at least 40–100 mb/s (40 for the more remote areas) with many fiber-connected homes. The technology for even simple copper-based lines (ADSL2+ and cable) already can do 20–100. And since Moore's law and all the other laws are still valid in 10 years, it will not be problem to go higher. Now at some time it will cost more than just put fiber to the homes, so we will go that way, providing almost unlimited bandwidth possibilities."
—**Egon Verharen**, innovation manager, SURFnet (Dutch national education and research network)

"Ninety percent of all Americans is too high a number for adoption in a decade. By 2014, there will be still be 25% of the American population that will not be connected to the Internet through their home using connections as fast as today's high-speed networks. This will be through a combination of factors that delay adoption, from persistently socioeconomically disenfranchised segments, less desirable markets for companies to target, to those that choose not to be connected." —**Dan Ness**, MetaFacts

"Satellite's too iffy. Telco's only needed to figure out a marketing/pricing strategy in the late '90s and couldn't. In that same time, cable tore up and repaved every road in America as they rolled out two-way networks. They won, and are not about to do it again to lay fiber." —**Steven Brier**, Brier Associates

> "This assumes a high degree of literacy and technical fluency in the U.S. population. With adult literacy rates among the lowest in the industrialized nations, I think it's unlikely."
>
> —Laura Breeden, Education Development Center

"Ninety percent of people won't be interested in going 'online' as we think of it now, though they may inadvertently be online, watching TV, or their refrigerator may send out for milk on its own. There are two groups: those who engage actively and those who do not. Of those that do go online, I'm sure 90% of those will do so on higher speed facilities." —**Sam Punnett**, FAD Research

ANONYMOUS COMMENTS

The following excellent contributions to this discussion of high-speed networks are from predictors who chose to remain anonymous.

"No way will we get 90% deployment of next-generation, high-speed connections in 10 years. That would imply a massively expensive infrastructure rollout."

"So far, the speed of networks has been increasing quite steadily. The demand is very far from being satisfied (today's home connections can't even support TV-quality video)."

"Dramatically faster, yes, but 90% of all Americans, I doubt that. Thirty percent of the kids won't even have access at a grand level in their schools. How will they have this in their homes?"

"'Dramatically' needs to be 1 gbps up and down. Yes, that's gigabit per second."

"I don't expect it to be that high of a percentage. I expect broadband to fully replace dial-up. I don't expect the total percentage of home users to grow much larger. It looks like it has reached a 'natural limit' determined by other factors, such as education, interests, etc."

"Costs will go down. And high-speed access will be a must to access the best content. The market will push people towards it."

"Ninety percent is a bit high, given our poverty rate. But maybe—TV managed to penetrate that far. So maybe."

"Not sure about the percentages, but yes, I suspect most Americans will enjoy such connections. The real concerns are about those who are left behind—people living on [American] Indian reservations, the homeless—who will be even less connected to the resulting culture."

"Speeds may be up, but dramatically? Throughput will rise more slowly than volume as what gets shipped grows faster than the media of shipping. And no more than 75% of Americans will have daily access, and an increasingly large number will be only able to access via third party (provided by someone, but with restrictions) and controlled public (sometimes at a public kiosk or library) on a more sporadic basis. That is, social-access rates will actually be far below network-access rates."

> "While you're looking for cheap predictions, here are some more: The CPUs will be faster, computers will have more memory, and there will be more computers."
>
> —Anonymous respondent

"There is too much old technology that will last much longer than 10 years."

"Increasing income disparity will continue to keep at least 20% from having Internet in the home."

"By 2014, most people will be 'online' everywhere they go using broadband spectrum connections when 'on the go' and fiber when at home. At least 2% of the population will be 'unconnected' whether they can afford it or not and a significant population (20–40%) will not be able to afford these connections."

> **"This is purely an economic limitation; it seems likely in the time frame indicated."**
>
> —Anonymous respondent

"By 2014, the fiber will be plastic, driving the costs down, and the fiber will come all the way to the home, rather than the current node architecture that exists today. It will be cheaper. It will be faster. But they'll also go online from anywhere, anytime—wirelessly."

"That's just not possible, given the small numbers still that are online today. Unless the government gives every home a computer and high-speed connections are as common as regular phones, it won't happen."

"It may be satellite instead by then."

"I agree that networks will be faster, but it seems unlikely that 90% of Americans will have access to this technology for socioeconomic reasons."

"It will take more than 10 years to get 90% of people on 'small broadband' (less than three Mbps) networks; we certainly won't see 90% on 'big broadband' networks."

"Not in 10 years. Even now, only 94% of the U.S. have phones."

"The trend has the right sign, but the proposed target is too ambitious."

"Broadband penetration will probably top out roughly where dial-up did—short of 90%. Today, several cities have 60–70% penetration rates. I suspect that we will see penetration rates in the low to mid-80s in most places."

"I don't expect broadband access to reach 90% of households."

"Today's broadband will most likely reach 90% penetration within 5 years. Following closely behind, 'big broadband' will roll out slowly at first, then accelerate. By 2014, the goal should be to have 100 megabits to one gigabit of connectivity to all homes, an achievable goal."

"The final-mile problem won't be solved to that degree."

"I agree, but more likely 60–80%. Many of those aged 60 (and older) today will not have changed their habits."

"The digital divide will persist so that 90% of Americans won't be online that soon."

"Infrastructure will not support this, unless satellite Internet technology improves."

"America lacks a commitment to infrastructure."

"The Bells will rush fiber to the home to recapture the telecom residential monopolies, and that will be that."

"[It will happen] via the electric grid."

"We know from the diffusion of other media technologies that to achieve 90% is not likely to happen in this time period. Many people will refuse to join the mass."

"This is a no-brainer. Cable Internet speeds are ramping up already from their original 1.5Mb/sec speed caps."

"The only dramatically faster option is last-mile fiber optics, which is very expensive, and for most people, does not offer sufficient gains to justify the cost. Even if it does, it will take 10 years to lay it, and no one has really started."

"This number can only be achieved if high-speed networks and Internet appliances are as inexpensive as today's low-end television sets. I know you are mostly concerned with Americans, but worldwide, this will absolutely not be true. People in the world today (outside North America) enjoy the Internet using 9600Kb transfer rates."

"The supply of new technology, mixed with lower costs, will drive high-speed Internet access."

"Unless you make cheap wireless networks that plug into people's homes, the economic stratification will not allow everyone to afford the homes that contain such networks. As with the advent of the cellular phone market any new type of device, a training or plug-and-play campaign will have to be put in place to make this happen, and I don't think that the government will support such a plan or help to put it in place. Aren't we already supposed to have high-definition TV (and maybe one day digital TV) as a standard form of broadcast?"

> **"I don't see this happening. There is not political 'traction' (or funding) for such a sweeping infrastructure change. I think the U.S. will continue to lag behind most other developed countries in this regard."**
>
> —Anonymous respondent

"Depending on the cost of access—if high-speed broadband continues to be sold through the oligopoly/local monopoly system, we won't see the number hit 90%. That is an extremely high figure, particularly when you consider that currently 90% of households aren't even online, let alone at high speed."

"Many people will be left behind, and broadband will become an upper-middle-class entitlement."

ENDNOTE

1. See "Broadband Penetration on the Upswing," published by the Pew Internet & American Life Project, April 19, 2004, available at http://www.pewInternet.org/PPF/r/121/report_display.asp.

PART 17

LOOKING BACK; LOOKING FORWARD

WHERE HAS THE INTERNET FALLEN SHORT OF EXPECTATIONS?

Since so many of the experts we contacted were early adopters of the Internet, we asked them to think back to their views a decade ago and assess where the use or impact of the Internet has fallen short of expectations.

Many experts are disappointed that spam and viruses have proliferated without check. The digital divide vexes quite a few experts. Many observe that education, health care, and civic life have not adopted the Internet as quickly as they had hoped. Others wish that download speeds were even faster and are looking forward to a "video Internet." And a number of experts said the Internet is just about as they had imagined it would be.

Here are some examples of the experts' thoughts on this question:

> "(1) Education—I thought distance learning would be more widespread. (2) Elections—I thought we would get to online

voting sooner. (3) E-commerce—I thought that online commerce would have a more devastating impact on local commerce and local taxation." —**Charles M. Firestone**, executive director of the Aspen Institute

"I did not expect that porn and objectionable content would have as large an impact as it has had on so many." —Anonymous respondent

"Politics still sucks. America's getting more totalitarian even as the populace is dancing in the streets to downloaded music." —Anonymous respondent

"As with radio, most of the hoped-for educational impact of the Internet didn't materialize." —**Simson L. Garfinkel**, Sandstorm Enterprises, an authority on computer security and columnist for *Technology Review*

"We forgot to build the Internet with enough security and economics." —Anonymous respondent

"As I feared, bland content from large media companies dominates too much. There is great creativity from a wide range of sources, and it does get noticed and it does have an impact. But the balance is not where I would like it to be."—Anonymous respondent

"It has exceeded my expectations for certain demographic segments of the world's population. As expected, most people in the world are unaffected by the advent of the Internet." —Anonymous respondent

Many respondents had been very generous with their time, and some were clearly growing tired of typing full sentences. One anonymous response: "Enhanced democracy: Not. Enriched sense of community: Not. Public space for learning: a mix but mostly commercial."

ADDITIONAL RESPONSES

Many other survey respondents shared incisive, overarching remarks to the question, "*Where has the Internet fallen short of expectations?*" Among them:

Looking Back; Looking Forward

"It hasn't." —**Moira K. Gunn**, host, *Tech Nation*

"The mainstream is slower to fully embrace it." —**James Brancheau**, VP, GartnerG2

"Civic participation. Compassion." —**Douglas Rushkoff**, author; New York University Interactive Telecommunications Program

"E-democracy. Virtual community." —**Barry Wellman**, University of Toronto

"Broadband has been slow to roll out, inhibiting the potential impact of true digital households." —**Mike Kelly**, America Online

"It has basically gone further than almost all of my expectations short of truly immersive experiences. But I have since decided that immersive virtual experiences are too dull compared to real experiences." —**Alexander Rose**, executive director, The Long Now Foundation (this organization works to promote long-term thinking)

"In generating civic participation, enlightened discourse, [and] more rational politics." —**Gary Chapman**, LBJ School of Public Affairs, University of Texas at Austin

"It has been less intellectually broadening than I would have hoped. The ready availability of alternative viewpoints has not made people more tolerant and broad-minded. Quite the opposite—it has, if anything, contributed to polarization, as technology has made it easier to seek out viewpoints that agree with your own preconceived notions, and to filter out views that might challenge those assumptions." —**Rich Jaroslovsky**, *Bloomberg News*, founding president, Online News Association

> "Civility online has deteriorated and the potential for online discourse as a means of revitalizing the public sphere has progressed more slowly than I had expected."
>
> —Howard Rheingold,
> Internet sociologist,
> writer, speaker

"The amount and take-up of services has exceeded almost all scholarly estimates of 1 or 2 decades ago." —**Mike Botein**, Media Center, New York Law School

"I expected it to have more of an impact on education than it has." —**Joe DeSantis**, Gingrich Communications

"Ten years ago, I expected that the Internet would change education in my lifetime." —**Christine Geith**, Michigan State University

"I know I expected to see far more online product-customization tools. I would have thought that by now everyone's clothes would all be made custom." —**Fred Hapgood**, Output Ltd., technology writer

> "Only that in the '90s it was more a vehicle to create open socities where it's now being used for surveillance and censorship because of the world circumstance."
>
> —Jonathan Peizer, CTO, Open Society Institute

"Spam and viruses. Identity theft and frauds abetted by easy anonymity and insufficient privacy safeguards. Dismal customer service at most companies—especially Internet businesses." —**Peter Denning**, Naval Postgraduate School, Monterey, California, columnist, *Communications of the ACM*

"As with radio, most of the hoped-for educational impact of the Internet didn't materialize. For information, see *Radio Research, McCarthyism and Paul F. Lazarsfeld*: http://www.simson.net/clips/academic/pfl_thesis_scan.pdf." —**Simson L. Garfinkel**, Sandstorm Enterprises, an authority on computer security and columnist for *Technology Review*

"I expected the Internet to completely revolutionize education (especially higher education) and other such institutions—government, news, etc. The technology evolved in a way that would have permitted this, but I underestimated the resistance of the institutions to change." —**Gary Bachula**, Internet2

"I expected that computers would be more available to families at all income levels. I expected less expensive access to proprietary databases. I did not anticipate the extent of intrusive activities such as spam, spyware, and unwanted data mining." —**Lois C. Ambash**, Metaforix Inc.

"Synchronous collaboration—'sharing minds' via shared electronic spaces." —**Noshir Contractor**, University of Illinois at Urbana-Champaign

"The growth of online multimedia news and interactive content has been hampered by slowness of adoption and other market forces, coupled with the 2001–2004 failure of the 'new economy' to measure up to advertisers' expectations." —**Mack Reed**, Digital Government Research Center, USC

"I am disappointed at the persistence of the digital divide and continued expense of broadband access. I am looking forward to increased access to online services to all members of society." —**Gary Kreps**, George Mason University, National Cancer Institute

"We still have a digital divide—so much of the country, of the world does not have access to the resources that many others enjoy." —**Brian Reich**, Internet strategist for Mindshare Interactive and editor of the political blog campaignwebreview.com

> "I'm disappointed that the Internet and new technologies have not strengthened the social fabric and increased civic behavior. I'm also concerned that the aggressive actions taken by entertainment-industry associations and the lobbyists to stop sharing are short-sighted and will harm the large entertainment companies ultimately more than protect them."
>
> —Terry Pittman, America Online, broadband division

"It has not fallen short of my expectations. It has exceeded them. I never expected to be spending 2 hours a day answering e-mail, most of it requiring an answer. It has improved my communications system beyond my imagination." —**Arlene Morgan**, Columbia Graduate School of Journalism

"I had thought it would do more to change us socially and culturally. I think we're still in the 'old wine in new bottles stage,' looking to use technologically transformed means of doing things to do the things we're used to. Linguists say that a pidgin language moves to the Creole stage (develops its own integrity and grammar) only in the second generation (since kids are better at language acquisition and system-grasping than adults). I think that's what will have to happen with the Internet. The so-called 'digital natives,' the generation coming of age, will find truly transformative ways of using the Internet." —**George Otte**, technology expert

"I thought the spread of the network would be faster, that it would be a feature in every home by now. The delay in the implementation of broadband was something that few predicted at the time." —**William B. Pickett**, Rose-Hulman Institute of Technology

"Access to online databases and publications has not become as widespread as quickly as I thought it would. Partly it is because of fear over protection of intellectual property. Partly it is because proprietary sources are charging too much, thus impeding growth." —**Ezra Miller**, Ibex Consulting, Ottawa, Canada

"The failure to address the security issues that were identified before the mass-market take-off. The predictions in 1995 that the take-off would be followed by a post-Y2K boom-bust and backlash has proved all too accurate. I was one of those who went public with such predictions (1996, 'The End is Nigh,' published by *Computer Weekly* and IMIS [Institute for Management of Computer Systems]) in the hope of seeing the necessary action to prevent them from happening. What happened was even worse than I had feared." —**Philip Virgo**, secretary general, EURIM

"I believed 10 years ago that the Internet would replace physical social networks, and would flatten them. What has happened instead is that society has found a way to make the Internet work in a way that supports a class society, rather than replace it. Now many people have access, but social rules around access have developed that keep social hierarchies in place. Collaboration has also fallen short of my expectations. It is still very difficult to collaborate online, and in the business world, the model of creating a document, attaching it to a message, and sending it is a replication of an old-world concept that makes no sense." —**Ted Eytan**, MD, Group Health Cooperative

"I see students increasingly assuming that if it is on the Web, then it is the complete and accurate story." —**Fran Hassencahl**, Old Dominion University

"It hasn't. It's right where it should be: The Internet is to the 2000s what TV was to the 1960s." —**Joshua Fouts**, executive director, USC Center on Public Diplomacy

"It took much longer for broadband to catch on in the U.S. and globally. Plus, independent online-only publications have mostly failed miserably to the old-line media companies online. Online gaming also took longer to become a mainstream passion than I thought it would." —**Mark Glaser**, Online Journalism Review, Online Publishers Association

"At the time, I thought we were all headed for a virtual reality world, but instead only a subset of users use MMOGs. I would not have predicted that the Internet would overcome TV watching among some users." —**Joe Crawford**, San Diego Blog, LAMP Host

"Convergence has been much slower than anticipated—ask Time Warner; roll out of broadband and digital television has been slower; computerish television—slower; the manner in which software has made products too vulnerable; the digital divide is still too wide; we have not adequately protected our children; we have allowed the loss of privacy; we do don't appreciate the power from the aggregation of information; some of the attempts at creating transaction-based markets have failed." —**Bradford C. Brown**, National Center for Technology and Law

> "Ten years ago, I believed we needed an AAA-like organization, a group to promote the rights of 'drivers' on the 'information highway.' I still believe that is true."
>
> —Kevin Taglang,
> private technology consultant

"In improving K–12 education and a global sense of possibility among young people. And in the expansion of projects like ThinkQuest and Global SchoolHouse—that should have grown exponentially and have frozen in place or declined." —**Cynthia Samuels**, Center for America Progress

"Student textbooks are not yet really online. Mainstream news organizations have failed to act creatively and nimbly to embrace the Net." —**Jan Schaffer**, J-Lab: The Institute for Interactive Journalism

"(1) Newspapers have done very little to make use of the unique multimedia and interactive properties of the Web.

(2) Online communities have not grown as quickly as I anticipated. (3) Publishers have been slow to take advantage of the interactive and targeting capabilities of the Web to offer new kinds of editorial services and useful information services, though they have been eager to exploit the target marketing capabilities." —**Janice Castro**, assistant dean, director, graduate journalism programs, Northwestern University

"We continue to support and promote products and services to the 'wired' and ignore those who are confused and unable to access the Web, such as the very poor and the elderly. (Example: Commerce Department's last report of the 'closing of the Digital Divide'—simply not true when only 20% of seniors use the Net vs. 80% of others. This is a wide wedge in society.)" —**Tobey Dichter**, Generations on Line, nonprofit Internet-literacy agency

"Civic engagement—through I would see more by government by now." —**Liz Rykert**, Meta Strategies Inc., Toronto, Canada

"Wireless/pervasive computing has arrived much more slowly than I'd expect. I continue to be amazed by the number of voluntary Internet exiles—people who are offline out of choice/lack of interest, not economic necessity. In my own area of research (e-politics/e-government), change has arrived more slowly than I would have imagined. Governments continue to talk a lot about 'breaking down silos' between vertically organized departments, but organizational change has been very slow to catch up to technological change, and without that organizational change, the capacity to provide innovative, single-window service remains limited. In the political arena, we've yet to see a significant victory that can be unambiguously attributed to the Internet. The [Minnesota

> "Formal education has changed less, much less than most futurists predicted. It is hard not to have been disappointed, to not have been suckered in, by the grandiose claims for rapid changes."
>
> —Douglas Levin, policy analyst, Cable in the Classroom

Governor Jesse] Ventura story seems to be partly a story about the Net, but there's too much disagreement to conclude that it was the decisive factor; and the [Democrat presidential candidate Howard] Dean story, while it demonstrated the impact of the Internet, did not produce ultimate victory. Net campaigners are still very much searching for the transformational models that will go beyond online fundraising and organizing; the Net is still more a tool for facilitating offline campaigning than a forum in itself (except for a relatively small number of digerati)." —**Alexandra Samuel**, Harvard University, Cairns Project (New York Law School)

> "I can't think of any (negatives or disappointments) because too much of my time is spent in pure amazement at what is possible with each new day."
>
> —Leonard Witt, PJNet.org

"In 1994 I suspected that within 10 years most people would be using e-mail, and that a great deal of commerce would be conducted online. I had little idea that broadband would allow the kinds of applications we have now. I did suspect more success in telework. This has moved more slowly than I expected it would." —**Alex Halavais**, State University of New York at Buffalo

"I thought the Internet would enable direct e-commerce from the producer/manufacturer to the consumer, bypassing the middleman. In part because of resistance by middlemen, this has generally not happened, Dell Computers notwithstanding. I thought that the widespread use of consumer-based online authentication systems would be here by now. I thought that the 'Local Wide Web' would be flourishing—in other words, using the Web to enhance local communities. It's only with 'Google Local' in the last few months that this is beginning to happen. I thought we'd be farther along on e-government, e-learning, e-health, e-transportation (e.g., ITS [intelligent transportation systems]) than we are now. I thought the Semantic Web would be here by now." —**Rob Atkinson**, Progressive Policy Institute (previously project director at the Congressional Office of Technology Assessment)

"I am disappointed with the extensive growth of proprietary programs and the lack of standards that permit interoperability."
—**Ted Christensen**, coordinator, Arizona Regents University, overseeing development of e-learning at Arizona's three public universities

"I think 10 years ago, most of us involved with the early Internet business models thought the Internet would replace traditional businesses. Instead, it has grown to enhance business in ways we could have never predicted." —**Gerard LaFond**, Persuasive Games, red TANGENT

"The use of the Net by educational institutions for high-quality course instruction and learning is in use and in continued development by those that embraced it, but there has been less use by traditional established institutions and governments than I expected." —**William Stewart**, LivingInternet.com

Anonymous Comments

The following excellent contributions to this discussion are from predictors who chose to remain anonymous.

"The means of finding what I want on the Internet have not developed nearly as much as I had hoped and expected. The flood of maliciousness and avarice that e-mail has unleashed is a big disappointment."

"Electronic government. Household tasks."

"K–12 education; health care; e-government; e-books."

"As I feared, bland content from large media companies dominates too much. There is great creativity from a wide range of sources, and it does get noticed and it does have an impact. But the balance is not where I would like it to be."

"Video—most video applications remain very crude. The network infrastructure simply isn't sufficient for reliable Internet video."

"Democratic life."

"Civic engagement and education."

"We were drastically wrong about predictions of online community."

"Fallen far short in electronic democracy or building political capital."

"The Internet has been bogged down by commercial Web sites that reflect the dominant corporations. I would have expected more 'shopping bots' and other consumer tools to become ubiquitous tools."

"Local governance."

"Fewer people with access and expertise than expected."

"I still struggle to control the flow of information so that I only engage when I want to, not when someone else wants to engage me."

> "Things are happening. Things are still poised to become radically different over time. But the virtual world still must be 'digested' by the real world. Just because it's there doesn't mean it's having the maximum impact possible on our lives."
>
> —Anonymous respondent

"The quality of Internet content continues to be generally poor. It is still the case that 'on the Internet nobody knows you're a dog,' and so one cannot trust Internet relationships."

"I had few expectations, but was always surprised at how long it took to realize what could easily be imagined. Standards wars seemed to hold things up."

"Eradicating porno sites, safeguarding computers against virus invasions, moving too fast and not thinking of rules/regulations for Web sites, so much spam. I must have 40 spam stops every day at work, and on my home Yahoo account I must get 300 a day. And I still get junk mail at home."

"Online learning has been very flat. There has been a lack of successful collaborations among institutions, which are negatively incented to form partnerships (due to enrollment quotas) and to use technology (faculty do not receive positive incentives to explore technology rather than publish papers)."

"Health care—especially in terms of the ability to exchange patient information. Education hasn't really changed fundamentally. Communities of place have not changed in the ways we hoped."

> "I thought that the Library of Congress would be digitized by now. I am appalled that this has not happened. The cost of such an undertaking would be about $230 million. I am disappointed, but not surprised that encryption technology isn't widely used today (most e-mail is not encrypted end-to-end, and neither is this survey)."
>
> —Anonymous respondent

"The future seems to always lag expectations. There was no 'Big Brother' in 1984 and no 'HAL' in 2001. I'm not surprised at where we are; I would have thought we would be here sooner. The places where we've fallen short have been in the video streaming area. I would have thought 10 years ago that nearly all entertainment would utilize the Internet for all forms of television, on-demand movies, etc. That hasn't happened yet."

"The utility of e-mail due to spam, the validness of information (due to people's ability to publish what they want, whether wrong or not), and weakness of the structure due to DoS attacks, co-opted machines by worms, etc."

"(1) Individual creation of video; (2) medicine and health care."

"Too many bandwidth-eating applications with all glitz and no substance."

"Disintermediation hasn't really happened. Consolidation of the pipeline and legal restriction of the potential of this pipeline was very sad to watch."

"There has been little or no change in the use, other than the explosion of online retailers. Virtual interaction has taken a backseat to advertising."

"Expected the Internet would hail an era of better customer satisfaction by making institutions more accessible. If anything, service has decreased as organizations have not been able to keep up with the demand."

"The Internet is very effective in increasing certain types of efficiency, e.g., in shopping. It has fallen short in differentiating quality information from everything else, which dominates."

"I was involved in some of the early 1990s conferences on health information infrastructure and telemedicine. Just finished reviewing some files tonight. Many of the same issues are still under debate with the same benefits being touted. We have, however, made progress with Internet access and broadband distribution. It's just much slower than anyone thought it would be. And the problems of electronic medical records and standards are taking generations to solve."

> "The whole thing is still way too slow, even with the best computers on the best network connections. It's still not reliable enough, and there is way too much garbage to wade through to find genuinely useful information."
>
> —Anonymous respondent

"I expected to get away from typing, to have functional voice recognition and easy graphical input. But we're not there. I'm still typing, and it hurts."

"The Internet is much less secure than I expected with the explosion of worms, viruses, spam, and fraud."

"Bots never really came to the fore."

"I had expected more change in the medical arena and an overall reduction of societal health care costs as a result of Internet communications and record-keeping efficiencies."

"I'd expected more innovative ways of seeing things, less 2D/screen, more immersive and engaging interfaces."

"I am still surprised that I am reading information on paper. I also thought, by now, 'on-demand' options—of movies, etc.—would be more pervasive. I still cannot make a doctor's appointment online."

"Micropayments have been a failure. The use in the arts is less than I would have expected."

"The Internet still does not provide easy access for most users. Sites are often designed for more highly educated, wealthier Americans. The Internet continues to be inaccessible on a variety of levels for many Americans."

"It has exceeded my expectations for certain demographic segments of the world's population. As expected, most people in the world are unaffected by the advent of the Internet."

"Religion. I launched faith-oriented Web sites in 1996, expecting many would see the Internet's effectiveness in religious communication. It didn't happen. Most adult churchgoers still can't use the Net effectively."

"Ten years ago, I would have predicted a higher penetration rate in homes, and a greater impact on health care."

"We haven't yet developed a method for ensuring the quality of information posted online, and there is a lot of junk out there that masquerades as useful knowledge."

"The growth in weak ties (if there indeed has been one) has not had the impact I imagined it would."

"I had expected the Internet to represent more diverse views, but the most popular sites are part of large media conglomerates or other large corporations."

> "Teleconferencing and computer-mediated real-time communication is the vision that has not happened. We still wait for the paperless office. The concept is probably foolish. Some folks (not I) thought that education would have been more rapidly transformed. I was a pessimist in this space."
>
> —Anonymous respondent

"It's about where I expected. I gave a talk at InternetWorld 2005, and mostly things have turned out as such. Web services are slow to get going."

"The public sector has embraced the Internet much more slowly than I had envisaged. Community networks have never really become mainstream as we thought they would.

Looking Back; Looking Forward 255

Also, digital libraries have mostly been driven by professional groups (such as ACM/IEEE), rather than by governments."

> "While I hoped it would be a big time saver, it has added a new demand to my personal and professional life. Being able to reach out to so many means you need to stay on top of them all. It also means more people can reach me."
>
> —Anonymous respondent

"Electronic money, i.e., a currency created on the Internet for electronic exchanges, unrelated to national currencies."

"Too few physicians using the Internet in practice, either to access information or to communicate with their patients."

"Ten years ago, I thought the Net would be a greater force for peace in the world. Now I see this as naive."

"Among other things, I thought journalists would grasp the technology's potential more firmly. We have not, for the most part, and we remain paralyzed by the threats we see from the changes."

"Small- and medium-sized businesses haven't participated in e-commerce as much as I would have expected."

"I expected video broadband delivery to be much further along."

"The Internet was supposed to empower individuals at the expense of corporations and nations. That view now seems naive."

"The 'cleanliness' of information available—I do not refer to pornography—I refer to the fact that there is so much 'dead' information that comes across daily when all I need is a statistic from one major source; I am disappointed by this."

"I had thought that we'd see more interlinking of various types of content by now (example: images of artworks linked to in-depth encyclopedias of art history) along with better visualization tools. The progress is being made but much more slowly than I had thought."

"The saddest part of the story is how big business and media has tried to force the Internet into the box they define for it, rather than using it to its fullest potential. Not that all is lost, but all of these analytics companies trying to measure and show business results is quite hideous. Also, the use of content online keeps getting downplayed when really that is what the Internet is about."

"It's taking longer for people to comprehend and embrace the power of the Internet than I thought would be the case. I'm surprised by how many people still don't 'get it.' I had also expected online education to be more widespread than it is right now."

"Effective personal instruction and creation of learning landscapes has been slow in coming. People outside of education made too many important decisions. Time of involvement, peer mentoring and sharing, and project-based learning and innovative ways of working seem to have been slow in coming, and just as they did, the No Child Left Behind Act created a tidal wave that swept over the technology initiatives in that there was confusion so the uses of the Internet were slowed."

"The Web is still largely a 'junk' medium, like a library where all the books are strewn on the floor (I stole that quote from someone but can't remember who). The Web has been co-opted by large media companies; evolution of the centralization of media has continued in a medium many hoped would discourage this problem. Web pages still load too slowly."

> "Fallen short? Quite the contrary! Due to the Internet, I telecommute 300 miles from my home office, and thus can be close to family rather than be stuck in another city due to job needs. I get and share information about uncommon things like treadle sewing machines, purchase parts that I would have enormous difficulty finding locally. Internet research allowed us to collect a depth and breadth of data about our neighborhood that we could never have gotten from a realtor and allowed us to find an underpriced house in a desirable neighborhood. The ability to do this sort of thing I would not have thought would have been possible in 1994."
>
> —Anonymous respondent

"Primarily in terms of the impact on politics and society in general. Like many people who have been researching the Net for some time, I also initially approached the Net with a somewhat utopian and deterministic approach. Research (and reality) have forced me to take a more balanced approach in terms of my expectations of the Net. In general, I see the Net as being a catalyst for existing social and political trends and changes rather than something that necessarily initiates these changes."

"The Internet still requires a great deal of human effort to filter through the 'noise' to find what one is looking for (i.e., there's still a lot of crap out there). I'm also surprised that extremely high-speed Internet access is not more widely available. In many places, dial-up is still the connection method of choice."

"The corporate buying and corralling of the Internet. There is a lot of information out there, but linking structures funnel you to only a small portion."

"The FCC vision of humans as consumers who consume third-party corporate content, as opposed to the Information Revolution where all online are creators and innovators. FCC policy has perpetuated the evolution from couch potato to cyber couch potato. Instead of promoting true Internet connections, the FCC has liberated a duopoly that has the power to ban VPNS, Web servers, Wi-Fi access points, encryption, third-party VoIP. FCC policy envisions the humans as 'consumers' of mass content who have no need for the empowerment of an interactive medium. Finally, FCC policy has led us to the place of mediocre broadband network access, provided by a consolidated market, with no concept of traditional common carriage ideals."

> "It has not brought people of all walks closer together. With the digital divide, there is a greater marginalization of certain groups. Too many lower strata families lack the buying power, computing power, and the overall knowledge power to function online."
>
> —Anonymous respondent

"I was surprised to see how poorly the IT sector deployed defenses to spam and hackers. I was surprised that Microsoft products have never improved in quality. Information visualization."

"I did not expect that porn and objectionable content would have as large an impact as it has had on so many."

"I never expected the Internet to take off like this. I would not have predicted the problems of the Internet—spam, phishing, viruses—and would have been disappointed to learn of them, if I had even thought they would happen."

"[I am disappointed in] its ability to redefine publishing and information-sharing models. Online publications have gained editorial legitimacy, but they are lagging in acceptance as a mass medium. And e-mail has become a near-nightmare. The inability of technocrats to fix the spam problem will go down as a major missed opportunity in Internet history."

"It has not become as ubiquitous as expected, and is not likely to rise over 75–80% penetration as long as we are tied to PCs as the main interface. I think that is overcome by widespread adoption of embedded devices, in everything from our cars to home appliances to medical devices such as blood-sugar and heart-rate monitors. Second digital media distribution. Both the technological and the business barriers have been far more difficult than anyone expected."

WHAT IMPACTS HAVE BEEN FELT MORE QUICKLY THAN EXPECTED?

Experts wrote with evident delight about the explosion of e-commerce, smart searches, mobile communication, and peer-to-peer file sharing. Others shared their disappointment that spam, identity theft, and other online pests have moved so quickly toward dominance. For example:

> "The astounding array of information available on the Internet is much larger than anyone could have ever expected."
> —Anonymous respondent

"The rise of the Web is astonishing. In 1992 (I have slides from a talk that year) we were not sure that the Web would win out over competitors such as WAIS, Archie, or Gopher. The transformation of the telephone industry has gone faster than I thought."
—Anonymous respondent

"E-mail was expected. The Web took us completely by surprise." —Anonymous respondent, a computing and Internet pioneer

"I don't subscribe to a newspaper anymore. I don't shop at retail stores nearly as often, or the bank. I don't buy reference books or go the library. I don't use the phone as much."

—Anonymous respondent

"I would never have imagined blogs, or that I would have one of my own. On the other hand, I spend much more time doing fairly routine work (such as scheduling meetings) online. The nuisances, like spam, viruses, and comment spam, are worse than I would have predicted." —**Peter Levine**, deputy director, Center for Information and Research on Civic Learning and Engagement, University of Maryland

ADDITIONAL CREDITED RESPONSES

Many other survey respondents shared incisive, overarching remarks to the question: *"What impacts have been felt more quickly than expected?"* Among them:

"Mobile Internet, both on cellphones and Wi-Fi. GPS and other services in cars. Online search capabilities (i.e., Googlization). Net-centric warfare (i.e., uses by the military on the ground). Pervasive impact of e-mail and instant messaging in our daily routines. Political organizing and coordinating by Internet."
—**Charles M. Firestone**, executive director of the Aspen Institute

"Free services. Ten years ago I would never have predicted that Yahoo would give away all the stuff it does. All those features were business models, once upon a time." —**Fred Hapgood**, Output Ltd., technology writer

"The rise of P2P networks exploded more quickly than I expected." —**Andy Opel**, PhD, Dept. of Communication, Florida State University

"Online gambling." —**Simson L. Garfinkel**, Sandstorm Enterprises, an authority on computer security and columnist for *Technology Review*

"Acceleration of market forces." —**Douglas Rushkoff**, author; New York University Interactive Telecommunications Program

"News media." —**Peter M. Shane**, author of "The Electronic Federalist," in the book *Democracy Online: The Prospects for Political Renewal Through the Internet* (2004)

"Extent and magnitude of adoption. Rise of new industries and businesses. Computing becoming a household word and concept." —**Peter Denning**, Naval Postgraduate School, Monterey, California, columnist, *Communications of the ACM*

"Work at home via attachments. Widespread networking via e-mail and IM, for both community and work." —**Barry Wellman**, University of Toronto

"Consolidation and advertising growth." —**Mike Kelly**, America Online

"The range of services and consumer take-up has been much greater than originally expected, apparently because the private sector has found ways—e.g., pop-up ads—of paying for things." —**Mike Botein**, Media Center, New York Law School

"Americans' use of e-mail in general. Americans' use of the Web as a major source of reference information." —**Daniel Z. Sands**, MD, MPH, Beth Israel Deaconess Medical Center, Harvard Medical School

"E-commerce—who would have predicted that Amazon would be the first successful e-tailer? Online auctions. Travel services. All deployed faster than I would have guessed, driven in large measure by consumer willingness to use credit cards online. Online dating services have been kind of a surprise to me, as well." —**J. Scott Marcus**, Federal Communications Commission

"Music sharing. Hope in politics. Customer-to-customer conversation subverting marketing." —**David Weinberger**, Evident Marketing Inc.

"Globalization of Internet access and use." —**Bill Eager**, Internet expert

"The complete ubiquity of spam (I now filter out about 95% of my e-mail to a spam bucket just to get to the e-mails sent to me by people I know or those who need to contact me for bonafide reasons); the wildfire-like spread of personal publishing via easy-to-use blogging software; the influence of political blogging on public opinion; the swift-rise of social-network models for creating, disseminating, and selling culture (e.g., music file sharing, resource sites such as http://istockphoto.com and art communities such as http://deviantart.com)." —**Mack Reed**, Digital Government Research Center, USC

> "The assumption, by the wired intelligentsia, that they can find out pretty much anything on Google."
>
> —Dan Froomking, washingtonpost.com, neimanwatchdog.com

"Political impacts, primarily due to blogging, have happened more quickly and more intensely than I expected. The prospect of everyone with a computer and an Internet connection becoming a publisher has been swifter and more pervasive than I expected. The adoption of Internet technologies by government agencies has been quicker than I anticipated. The slow death of newspapers has not been as slow as I had hoped." —**Lois C. Ambash**, Metaforix Inc.

"The ubiquity of the Internet in daily life—from making travel reservations to looking up movie times. The dependence on the Internet to do virtually every activity was unforeseen by me." —**Ted Eytan**, MD, Group Health Cooperative

"The commoditization of content has occurred faster than I expected. The majority of online people believe that most of the news and information they need is available free online." —**Terry Pittman**, America Online, broadband division

"The emergence of digital media channels, and the use of the Internet as an entertainment medium." —**Brenda Hodgson**, Hill & Knowlton

"Adoption of online commerce by businesses has happened more quickly than I had imagined." —**Gary Kreps**, George Mason University, National Cancer Institute

"Its impact on information exchange; I did not expect so much information (from the mundane to the highly technical) to be online and so freely accessible." —**David Tewksbury**, University of Illinois at Urbana-Champaign

"The embedding of the Net into the very fabric of life, at least for those who are connected, has been more dramatic than I'd expected so fast." —**Dan Gillmor**, technology columnist, San Jose Mercury News, author of *We, the Media*

"I never expected it to have the kind of penetration it does now. By 1994, I was already predicting that the Internet would follow the same path other technologies did, and that there would be a consolidation of producers. I've been surprised by the number of people who produce their own content for the Web, and by the degree to which peer-to-peer communication occurs." —**Alex Halavais**, State University of New York at Buffalo

> "Mobile telephones, PCs, and the Internet have untethered computation from the desktop and colonized all parts of our lives faster than I had expected."
>
> —Howard Rheingold, Internet sociologist, writer, speaker

"Greater penetration to remote parts of the world, and adoption in places where access is limited." —**Joshua Fouts**, executive director, USC Center on Public Diplomacy

"Ubiquity, internationalization, pollution." —**Louis Pouzin**, Internet pioneer, inventor of "Datagram" networking

"The transformation of intellectual property is, I think, already a done deal—it will just take a little while for our laws and policies to catch up. And as a social phenomenon, people have very quickly acculturated to the concept of 24/7 information availability." —**Alexandra Samuel**, Harvard University, Cairns Project (New York Law School)

"IP phones. PDA and wireless applications." —**Liz Rykert**, Meta Strategies Inc., Toronto, Canada

"The e-mail system has turned out to be a lot more useful than I expected allowing instant communication with research fel-

lows abroad. It has also been extremely useful to be able to access libraries and other knowledge banks from distance."
—**Kirsten Mogensen**, Roskilde University, Denmark

"The alternative media—illegal file downloaders, bloggers challenging CBS or Trent Lott. Dissonance always has a place in American society, but as people try to control all aspects of a culture, be it entertainment or journalism, the Internet can more easily spark an opposition and have it spread almost uncontrollably." —**Brian Reich**, Internet strategist for Mindshare Interactive and editor of the political blog, campaignwebreview.com

"Yes, the Internet saves time and makes communications easy, but the increased communications resulting from the Internet means that more time is devoted to it." —**David M. Scott**, Freshspot Marketing, *EContent* magazine

> "Political activism has been striking. Disenfranchisement of mainstream media is happening more quickly than I would have thought."
>
> —Jan Schaffer,
> J-Lab: The Institute
> for Interactive Journalism

"I would never have imagined blogs, or that I would have one of my own. On the other hand, I spend much more time doing fairly routine work (such as scheduling meetings) online. The nuisances, like spam, viruses, and comment spam, are worse than I would have predicted."
—**Peter Levine**, deputy director, Center for Information and Research on Civic Learning and Engagement, University of Maryland

"As a tool to conduct transactions and as a pipeline for global trade; the integration with PDAs; the reliance as a communications tool; the transition to getting children to learn computing at younger and younger ages." —**Bradford C. Brown**, National Center for Technology and Law

"The pace of criminal exploitation; once there were sufficient pickings for the Internet to be taken seriously as a source of revenue." —**Philip Virgo**, secretary general, EURIM

"I am amazed at the spread and rate of access. I think everyone is. We are discovering that virtually everyone coming to college has access, is adept. Most faculty have scarcely begun to take advantage of that—may not be able to, in fact, may have to wait for the next generation to do that. But the students are already there, online; access is already approaching the saturation point, and that raises critical questions about what happens then." —**George Otte**, technology expert

"It may sound silly, but the shopping power of the Net has changed how I do business. It has really been a real time-saver for me, and I think for many women." —**Arlene Morgan**, Columbia Graduate School of Journalism

"The rise of Weblogs as influencers in the media and public." —**Mark Glaser**, *Online Journalism Review*, Online Publishers Association

"The movement of the Internet into wireless and cellular communications; the prosperity of companies like eBay and Amazon; and the use of the Internet for free search: Google." —**William B. Pickett**, Rose-Hulman Institute of Technology

"The decline in national television audiences for broadcast news is more rapid than I expected." —**Janice Castro**, assistant dean, director, graduate journalism programs, Northwestern University

"The spread of rich media has exceeded my expectations, as has the quality and availability of wireless Internet service. I am sitting in front of my TV with my wireless-enabled laptop tapping this out at my leisure, without having to be secluded in my lonely basement office with my desktop computer. Earlier tonight, I checked in and watched a segment of *NOW* with Bill Moyers that I missed when it aired on TV, and the video quality, though coming through on a picture frame that probably measured only three inches by three inches, was great, as was the sound. So the distribution of quality wireless audio and video has exceeded my expectations." —**Kevin Featherly**, news editor, *Healthcare Informatics*, McGraw-Hill Healthcare Publishing

> "The Boomers are on board. They're using it and loving it. Who predicted their 70-year-old mother would buy an iMac, then learn to use it so she could get pics of her granddaughter by e-mail? If it gives them an emotional payback, they'll make the investment in time and resources."
>
> —Meg Houston Maker, user experience designer

"Those affecting daily activities (such as buying books, music, shopping for travel, checking the weather, getting driving directions)." —**Douglas Levin**, policy analyst, Cable in the Classroom

"The rearrangement of priorities in the media industries and the high level of confusion there have been happening pretty fast. And the success of open software as a technology and as a movement has been a happy surprise for me; never would have expected it 10 years ago." —**Tom Streeter**, University of Vermont

> "I really didn't expect the War Against the RIAA (aka, The RIAA vs. Damn Near Everybody). That one, I just did not see coming."
>
> —Mike O'Brien, The Aerospace Corporation

"The expansion of wireless access has made the Internet far more pervasive that I could have ever imagined. It's not about going to a computer in the home or office. It's about checking e-mail or driving directions while waiting for a cab on the sidewalk." —**Allen Fuller**, Fleishman-Hillard

"Everyday people have embraced technology faster than I expected—not just technology buffs or white-collar workers." —**Dan Ness**, MetaFacts

"I never expected to experience the dreaded feeling of isolation that overwhelms me if my Internet connection (particularly e-mail) is down for more than half a day." —**Peter W. Van Ness**, principal, Van Ness Group

"The use of e-mail, and the way it has radically changed the workplace. I can now do most of my job via a networked PC, and send and receive e-mails on the move (via BlackBerry… an amazing device!). Just a decade ago I was still writing paper memos and reports. Yes, face-to-face communications are vitally important, but the productivity gains from being able to 'copy in' colleagues, and forward ideas, are incalculable." —**Graham Lovelace**, Lovelacemedia Ltd., U.K.

"The radical change and acceptance in the way we live and work." —**Tiffany Shlain**, founder, the Webby Awards

Anonymous Comments

The following excellent contributions to this discussion are from predictors who chose to remain anonymous.

"Bandwidth increased more than my pessimistic projection due to competition. However, I predicted that this was the only way to get the higher speed."

"Weblogs and peer-to-peer networks."

"Organized crime has embraced the Internet much more rapidly than I had envisaged. This has resulted in the proliferation of spam, viruses, worms, pornography, etc. Overall, the younger generation has embraced online technology more rapidly than I anticipated."

"Online auction and online sales is much further along."

"Spam/identity theft."

"The pervasiveness and use of e-mail as a frequent substitute for telephone; high quality 'smart' searching does a better job than I had expected, making all sorts of information available without going to a library or calling a government office."

> "People's ability to find precise information quickly and accurately is one of the reasons for productivity growth. This has happened more quickly than I expected and seems likely to continue more quickly than economists expect."
>
> —Anonymous respondent

"Commercialization, commodification in general were something I never expected at this scale. I was a doubter of online auctions as well, but am now an avid eBay user."

"P2P file sharing, especially music; ubiquity of search engines; spam; easy-to-use viruses; other hacking tools."

"The way we have become dependent on the Internet as an information resource and on e-mail as an interactive communication tool."

Looking Back; Looking Forward

"The digitization of media and its impact on copyright and many businesses snuck up on me."

"Mostly in the form of entertainment and information available, especially to scholars."

"The quality and quantity of information in all fields available via the Internet."

> "Copyright and intellectual property battles have emerged much faster than I would have predicted, and seem to be biased in favor of copyright holders (something I would not have predicted)."
>
> —Anonymous respondent

"My work has been transformed by the Internet. The magazine I edit couldn't exist if it weren't for the Internet and couldn't publish the range of content and contributors it does with the Internet."

"It's had a huge impact on work life. I now spend my day in front of the computer using the Internet and most of my colleagues at my think tank do the same."

"The Web has really transformed the Internet—I couldn't have foreseen that 10 years ago."

"Everybody uses e-mail. The biggest surprise is its adoption by senior citizens."

"Copyright issues and fair use."

"The revolution in content distribution. The potential was obvious, but I felt resistance by the 'powers that be' (RIAA, etc.) would be more effective. Their moronic handling of the file-sharing dilemma accelerated the growth of P2P."

"Changing of business models such as music/CDs and the jones felt by users when the Internet is 'broken'/not available."

"Communications among family members of all ages, relationships, and generations. Changing the dynamics of education, at both high school and college levels."

"I'm surprised at how quickly e-commerce has taken off. The social habit of shopping."

"Its ubiquity has grown faster than I expected."

"Its influence in the supplying industries and the rate of change in the supply industry."

"Those astute in the use of the new tools have made huge leaps in their daily lives, in their understanding of the world, even in their financial fortune. It is the dawn of a new era."

"Connectivity. The rise of IMs and enhanced e-mail has helped people connect. Gaming has affected the ways we look at entertainment—accelerated the shift toward interactivity."

"E-mail has come to dominate daily life in a way I'd never have imagined."

> "The incredible abundance of information of any quality. Certainly, humans can comprehend and assimilate only a finite quantity, and there does not seem to be a limit in sight. That begs the question: Is too much unfiltered information really desirable?"
>
> —Anonymous respondent

"Voice over IP seems to be catching on quite quickly. The effect on the quality of film and television (downward trend in production standards) has been striking. The Internet is partly responsible."

"Ability to buy groceries without ever talking to a stranger or driving. (Did I mention I am an introvert?)"

"High-speed Internet has allowed huge attachments of files—beyond what was anticipated."

"Worrying about virus invasions."

"Commercial necessity of the Internet."

"The successful chilling effect of insane protectionist lawsuits."

"Hijacking the Net with porn and other dubious enterprises."

"E-mail has changed our lives—communication is now instant and written."

"Acceptance of technology has been quicker than I would have thought. Integration of hardware/software into daily routines has occurred at a much higher level than I thought it would."

"Extent to which print media adapted to the new format."

"Connectivity—IM."

"The drive for personal- rather than work-related need for equipment—new computers, printers, mobile devices—continues to surprise me. The unquestioned assumption of access, availability, being on the 7/24 clock also surprises me. Even though the seeds of this were there 10 years ago, the 'channel switching' needed to handle multiple roles and multiple social worlds all through the same devices is increasing much faster than I expected."

> "Spam is everywhere. I had no idea of the size of the spam problem. I am also surprised about the lack of a firm response on this issue. I also underestimated the magnitude of the security and virus problems that people have encountered. This is probably because I work on a Macintosh rather than on Windows."
>
> —Anonymous respondent

"Commercialization and the exiling of the wonderful geek culture far into the sidelines."

"The deep penetration of e-mail into every facet of daily life."

"The ability of the bad guys—see Osama and his ilk—of using the Internet as a soapbox."

"The surge to online usage in the mid- to late 1990s clearly spiked and came faster than envisioned."

"Access to more information than expected. Search tools. Comparative tools. Advertising and spam—both worse than feared."

"Communication via the Internet—whether peer-to-peer or peer-to-many, B2B, or B2C—has become a default for most people."

"Government has more functioning Web sites that provide useful information, but we still have a long way to go."

"It's become a major mass-advertising and retail medium. I probably would have predicted that B2B would grow more first."

"Changes in distribution of news have been quicker or more extensive than I had thought possible. Most folk that I talk with get their breaking news on the Internet and read newspapers and magazines for analysis of that news. Television is virtually out of the picture except for documentaries."

"I underestimated its take-up by lay users."

"I did not expect that file sharing would take off so dramatically. I did not expect that blogging, the independent journalist, would attract so much attention."

"Cell phone use in entrepreneurial and personal communications in underdeveloped nations."

"I underestimated its monstrous effect on political discourse."

> "The increase and bursting of the dot-com bubble happened relatively more quickly—people should have had more realistic expectations that when you are significantly challenging so many societal norms (vertical to horizontal, rewriting intellectual property rules) that it will take some time. No one was prepared for the speed of repricing (commoditization) of info I think."
>
> —Anonymous respondent

"Change to the work environment—reliance and predominance of e-mail in communications."

"The electronic nature of so many relationships is surprising (and not unwelcome) to me. Even now, you can live online."

"The closure of small, used bookstores."

"Social exclusion. So few of the world's population are part of the Internet, and one-third of Americans are not. There seems to be no real enthusiasm or plan to do something about this."

"E-commerce. Despite the dot-com bust, it is becoming hard to find even small businesses without a Web presence, and the number of them with e-stores is amazing."

"The prevalence of online shopping for all kinds of things."

"The dot-com boom made online financial transactions and Web-based transactions everywhere more standard than I would have expected."

"I am surprised by how the Internet has cut across generations, meaning my parents are using it (perhaps not as widely), as are a number of individuals in the 70+ bracket."

> **"People taking over the diagnosis of their own health issues; taking the power of control out of the doctors' hands and into their own. It was a century-old power relationship that was turned on its head in a few years."**
>
> —Anonymous respondent

"The explosion of the Web, providing a common interface not controlled (yet) by any one organization."

"Tremendous strides have been made in geo-spatial resources on the Web."

"The speed my kids are embracing new technology blows my mind; it's hard to imagine where their children will take us."

"P2P networks, blogs, and a few others are reshaping entire industries."

"Peer-to-peer greatly impacted/rattled the establishment. The acceleration of broadband access has definitely opened a lot of doors that were previously cyber-shut. It's improved the access to the marketplace for small business/entrepreneurs."

"The availability of information has really surprised me. For example, I Googled 'R/2R' and immediately got data on a chip made by Bournes."

"The cultural polarization due to the ease with which people can selectively filter what they see and read."

"Utter and ill-advised reliance on dubious information received via the Net; the blurring of the line between work and leisure; the movement of information online."

"Corporate data harvesting. The only thing saving us is that people have not really figured out how to use all that data we are collecting. But give it time."

"The culture has become homogenized by television and the Internet. It has broken down many barriers but made us aware we are more the same than distinct from each other. Otherwise, the speed of communication has made us realize how small the world is. Yet we still wish to travel and meet other people for the need of face-to-face, human interaction."

"Pervasive use of e-mail, RSS, IM, and blogging."

"Adoption of instant messaging still fascinates me."

"The impacts of blogs and online news sources lately are accelerating faster than I originally anticipated."

"The proliferation of social uses and media formats, e.g., with P2P file sharing."

"I don't subscribe to a newspaper anymore. I don't shop at retail stores nearly as often, or the bank. I don't buy reference books or go the library. I don't use the phone as much."

> "Computer-assisted pedagogies have turned out to be largely a disappointment to me, but the 'Google-Library of Babel' is so beautiful it still takes my breath away. I feel like I am in Alexandria, and I'm impatient for all works to be accessible in this way."
>
> —Anonymous respondent

"The disruptive effect on the media."

"Commercialization. It astounds me how fast the previously money-free elements of the online world have been subsumed by for-profit entities. Anyone remember InterNIC or pre-blog personal Web sites?"

"The uptake of the Internet by the general public has been astonishingly rapid, most especially the Web. I would have expected uptake to be primarily a generational phenomenon, but clearly there are a lot of middle-aged folks who are vigorously using the Internet now, who weren't aware of it 10 years ago."

"I am amazed how quickly the computer/Internet has become a must-have tool for families and stay-at-home parents. It replaces the newspaper and the phone book, it gives directions, it helps neighbors communicate about everyday items, and more."

"The impact of open-source software, and the ability of users to become very knowledgeable about the Internet."

"The commercialization of the Internet and its use as a political force."

"Searchable information via Google."

"Commerce is moving very fast-paced. Governments are also being impacted significantly."

"Internet adoption by older consumers, the change in retailing, the sheer number of providers of all types that one can interact with over the Web."

"Increased expectations of work productivity. Need to remain 'tech savvy' to do nontechnical work."

"The move from print on paper to digitization of scholarly materials. My students think that all information of any value is located on the computer—and they don't go to libraries anymore."

"Institutions and individuals have put far more information up than I expected, and much more quickly. This treasure trove, much of which is freely available, is a major attraction in drawing people onto the Internet. Commercial services (especially for travel and financial services) have developed quite well, and provide real value."

> "The impact on our political system and the availability of information from all levels of government."
>
> —Anonymous respondent

"The ability to retrieve 'general opinions' about issues; different from 'grounded research,' it is possible to find out what anyone thinks about anything ranging from fixing the toilet to diagnosing schizophrenia. I'm continually amazed at how misinformed people publish misinformed information that further misinforms others who read the misinformation."

What's Next?

We ended the survey with the most forward-looking questions: *"What are you anxious to see happen? What is your dream application, or where would you hope to see the most path-breaking developments in the next decade?"*

Mike O'Brien, a computer scientist with The Aerospace Corporation, wrote, "Fully immersive 3D alternate reality, portable. A completely separate and completely virtual world, equally accessible wherever or whenever you are. It would give a mental 'face' to the Internet that would allow people to get a visceral handle on it. Right now, the average Joe's vision of the Internet is like the blind men and the elephant—people think that what they see and use every day is the whole thing."

One expert wrote, "Artificial intelligences in appliances, vehicles, computer software. For example, I'd love a word processor that worked like a great copy editor—not simply a spell-checker or simple grammar checker. Or a kitchen appliance that would read all the bar codes of items in my pantry and refrigerator and recommend innovative menus, remind about expiration dates and calculate nutritional values for meals. Perhaps it would even use avatars to walk through recipes. Or, if activated, I'd like such a device to answer a question like, 'Where are the kids right now?' If each child has a cell phone (or PDA-type device) my kitchen appliance would tell me that Mary is at her friend's house and that Tommy is in the park. It might ask if I want to send them a note reminding them to be home by 5 p.m. to get ready for dinner. These are the sorts of network applications that enhance but also transform."

One respondent wrote, "Converged devices are a dream. I would love one phone/PDA that can get 2–4 lines, do e-mail (GPRS and real time), Wi-Fi, has Blue-tooth, IM, and video—and fits in my shirt pocket and does not cost more than $300."

One respondent wrote, "I would like to have the data about me in a virtual passport that I control and that I can choose who is allowed to see what specific information I choose within that passport. I would like to have my home—the appliances, lights, vehicles wired and knowing me and my preferences. I am interested in how nanotechnology is going to impact the products we buy today, the health care advances that we will be able to see, and the new products that will be created through nanotech applications."

Lois C. Ambash, president of Metaforix Inc. wrote, "My dream application is a fail-safe, user-controlled, user-friendly privacy screen that would allow people to reap all the benefits of cyberspace with none of the personal risks. What I am most anxious to see is genuine conversation between geeks and newbies. Many people who could reap great benefits from the Internet are hampered by the jargon barrier (and other language barriers, such as reading level and lack of facility in English). My dream situation—as opposed to application—will occur when beta testing routinely requires that any intelligent adult be able to use the product or application competently without a geek in the family or a lengthy interaction with tech support."

Additional Credited Responses

Many other survey respondents shared incisive, overarching remarks to the question, *"What are you anxious to see happen?"* Among them:

> "I would like to see secure online voting available to all. I'm anxious for 3G (or more) mobile broadband Internet access and telemedicine, especially in application to public health throughout the world." —**Charles M. Firestone**, executive director of the Aspen Institute

> "I am most anxious to see the keepers of our intellectual property laws admit that they are failing and restructure them in a smarter way. I am most anxious for a seamless open-source, online computer, meaning I buy something at an electronics store that when I plug it into the Net updates itself completely and keeps itself up to date in terms of operating system, e-mail, and Web clients all with open-source applications."
> —**Alexander Rose**, executive director, The Long Now Foundation

"Ease-of-use in deploying and integrating interoperable modules of open-source applications that can be used effectively by organizations without the money to buy experienced technical support. It is still too hard for most groups to take advantage of the tools already available. Computer science should figure out a way to build a Web-oriented software architecture that allows interoperability, modularization, reuse of code, and most importantly ease of use." —**Gary Chapman**, LBJ School of Public Affairs, University of Texas at Austin

"I want to have TiVo apply to all aspects of my life." —**Jonathan Band**, an attorney specializing in e-commerce and intellectual property with Morrison and Foerster LLP in Washington

"I am very interested in—but very uncertain about—the future of virtual or synthetic worlds. It would be exciting if they were to become an important social landscape, a place where people went to school, held social events, and the like. Increasingly virtual worlds are the only corners of life left where people are free to create and build as they like, without worrying about fire codes and licenses. They are a kind of utopia in that respect, and I hope they cast a long shadow." —**Fred Hapgood**, Output Ltd., technology writer

"Ways of seeing my groups (social network) that aren't boneheaded, stupid, and that don't require me to explicitly reconstitute them. A big deal: Integration of virtual networks with geography-based neighborhoods. This will be enabled by the integration of GIS." —**David Weinberger**, Evident Marketing Inc.

"Convergence isn't a device but digital content and data finding its path to the consumer in whatever way they want to receive it." —**Mike Kelly**, America Online

"Barring some type of massive infrastructure investment, we should concentrate on expanding the availability of hardware and networks (i.e., the traditional 'digital divide' problem)." —**Mike Botein**, Media Center, New York Law School

"I would like to see the ability to transmit very, very high-definition 'virtual reality' over the Internet so that users could 'experience' remote worlds, whether they be a trip up the Amazon, or to New Zealand, or to a completely imaginary world created by an artist." —**Gary Bachula**, Internet2

"Software that is trustworthy, dependable, reliable, usable, and safe and contains enough self-diagnostics that users can ask it what is the matter." —**Peter Denning**, Naval Postgraduate School, Monterey, California, columnist, *Communications of the ACM*

"I am anxious to see art and design enter into our devices, interfaces, and applications. The tools we use should be as beautiful as the art objects in museums of modern art." —**Christine Geith**, Michigan State University

> "The biggest issue is one of control. Forces of centralization are trying to control what we do online and how we do it. If they succeed, the promise of the Net will fade into a dim shadow of what could have been. I assume great technological developments. But people who care about freedom need to remember that policy is decided by activists, and they'll need to work much harder on the political side of things."
>
> —Dan Gillmor, technology columnist, San Jose Mercury News, author of *We, the Media*

"Dream application—project collaboration ability—to streamline the work of teams online in real time, having access to all of the knowledge contained within, assistance with planning, and implementation without fragmentation across multiple disparate applications. In health care, advanced medical vocabulary and contextual mapping that provide patients with information that is relevant to their care and personal situation, empowering them to make the best health care decision at any moment." —**Ted Eytan**, MD, Group Health Cooperative

"The dream application of the Internet is actually integration of several self-contained applications including VOIP, portable access, workgroup collaboration, GPS, and broadband video so that an individual can use a portable device to instantly access whatever or whomever they need." —**Bill Eager**, Internet expert

"I want to see online newspapers (and their partners) create virtual representations of real neighborhoods to help people lead more productive, sociable, charitable, safer, happier lives. I want to see traditional journalistic values prosper and spread on the Internet, but that requires newspapers to adapt to the new medium much faster than they've done so far." —Dan Froomkin, washingtonpost.com, niemanwatchdog.com

"I'm anxious about surveillance and lack of privacy. I'm also anxious about the average user's lack of understanding regarding the nuances of their use (e.g., the extent to which companies follow users' actions, etc.). I hope we can regain e-mail by figuring out ways to fight spam. Again, an important aspect is educating users about the system and how not to compromise their and others' personal information online." —**Eszter Hargittai**, Northwestern University

> "What has to happen for people to feel they have participated in a process in a meaningful way? How can a million people participate in something and each feel they made a contribution, a difference? What role could many-to-many play in world peace?"
>
> —Timothy L. Hansen, MoveOn.org

"I hate to admit it, but I don't really have a dream application. I have been involved with computers since 1952 when I wrote my first program for Whirlwind I. I am continually in awe of what is coming along." —**Bill Eccles**, Rose-Hulman Institute of Technology

"The ability to easily publish/share/access anything digital from any connected device." —**Terry Pittman**, America Online, broadband division

"Ubiquitous Wi-Fi coverage is a must. Convergence of Wi-Fi, home Internet service, VoIP, phone, and cell service—necessary in terms of pricing and service integration." —**Alexandra Samuel**, Harvard University, Cairns Project (New York Law School)

"Widespread civic literacy—people who understand how to use ICT to improve their face-to-face communities. A new model for cultural production, with millions or billions of producers, instead of a few mega-disinfotainment corporations." —**Howard Rheingold**, Internet sociologist, writer, speaker

"My 'most-wanted' applications: (1) Home-fabrication equipment. I want to be able to download the latest consumer device and 'print' it at home. (2) Lifelogging. Complete records of our lives. 'Reality TV' that allows you to re-experience others' experiences, either in real time or delayed. (3) Networked-based AI: technologies that find useful patterns in the Web at large and leverage these to provide for interesting connections between people and ideas. (4) An increase in freely available media through initiatives like the Creative Commons. I think this is particularly important when it comes to scientific literature." —**Alex Halavais**, State University of New York at Buffalo

"I would love to see true convergence—getting rid of all the many devices and their adapters to charge them up—and shift our technological innovation away from helping us to accomplish the obvious with respect to ICT and instead have it targeted at some of our most persistent problems—homelessness, HIV/AIDS in the developing world, a world free of violence. I think the arrival of the Internet has meant a fun and creative free-for-all as we zoomed in to figure out all the fun and useful things we could do. In 10 years, we will have figured that out, so I hope all that same energy will get targeted to make this world a better place." —**Liz Rykert**, Meta Strategies Inc., Toronto, Canada

> "Virtual reality via direct access to the brain. (Seeing without the use of your eyes, for example.)"
>
> —Jeffrey Boase,
> University of Toronto,
> Harvard University

"A vast improvement in the ability to filter out noise and focus on useful information. We need applications that reduce the quantity of information delivered to us while increasing its value and relevance. Such information filters must be permeable enough to allow through a user-definable level of divergent views, jarring notes, and out-of-left-field ideas and news." —**Rose Vines**, freelance technology journalist, *Australian PC User, Sydney Morning Herald*

"Network-controlled automobiles to reduce traffic congestion and accident rates. E-democracy tools that can ensure accurate, hack-proof voting and encourage nonparticipants to join in the democratic process so that representation will be fairer. Better voice-recognition and emulation tools and interfaces to make text-to-speech translation and keyboardless navigation more seamless. Artificial intelligence agents that can sort fact from opinion and political speech intuitively, and serve up completely objective accounts of politicians' activities and speech, so voters can make more-informed and less confused choices. A payment system that allows artists to be adequately compensated for their work by patrons who are, in turn, allowed to share culture freely. I could go on, but I need to get back to the 14 windows I have open on my screen right now." —Mack Reed, Digital Government Research Center, USC

"My dream application would be high-quality news sources for local communities, created by the people in those communities." —Peter Levine, deputy director, Center for Information and Research on Civic Learning and Engagement, University of Maryland

"I would like to see the Internet become more customized, and I don't mean that we choose all of our preferences; rather, the technology would be 'smart' enough to recognize our needs and wants and profiles and display information relative to that. Almost like a *Minority Report*-like thing (e.g., the billboards that recognize you and say 'you bought these pants yesterday'), but perhaps not that advanced. The ability to interact with information physically and spatially with our hands on large screens and surfaces."
—**Donna Tedesco**, Fidelity Investments

> "I would like to see more people have access to the Internet, particularly in the underdeveloped countries. Enhanced information access is the route to empowerment. I think it's important to come up with a spam solution. If not, spam traffic is going to kill the Internet. I would like to see a major funding project put in place that would allow anyone to access any publication in the Library of Congress...the actual contents, including illustrations."
>
> —Robert Lunn,
> FocalPoint Analytics,
> senior researcher,
> 2004 USC Digital Future Project

"Library resources. I'd like to see the creation of a legitimate academic library with vast resources available online. I'd like to see the rare scholarly journals, as well as the common ones, come online. I'd like to see communities of commentators form around works of literature, art, and scholarship in the kind of stable and secure environment that libraries provide. By libraries I mean nonprofit institutions that have stable funding for decades on end in the manner of major universities and state libraries in many parts of the world. I also mean institutions run by professionals who filter the outpouring of 'published' material according to accepted standards, so that users have a reasonable assurance that what they are getting is serious and creditable work." —**Stanley Chodorow**, University of California–San Diego, Council on Library and Information Resources

"I hope that we can use the Internet to make health care more productive and drive down the costs so that more Americans can obtain health care insurance." —**Bradford C. Brown**, National Center for Technology and Law

"I am anxious to see media companies break through their old way of thinking and start considering their readers as collaborators, as group fact-checkers, as people who are interested in making sure the whole story is told." —**Mark Glaser**, Online Journalism Review, Online Publishers Association

> "I am hoping that Tim Berners-Lee is ultimately successful in developing a semantic Web that understands that connections between disparate data in ways that really can make the flow of information as crucial a part of people's everyday experience as remembering to grab the car keys and wallet. If he pulls off his dream, the Internet will become a kind of universal thinking machine, and that prospect is terrific."
>
> —Kevin Featherly,
> news editor,
> *Healthcare Informatics*,
> McGraw-Hill Healthcare Publishing

"I would like to see the rise of a consensus that the nation, indeed, mankind, has an interest in the free, unimpeded growth of this means of human interaction. This means that both the not-for-profit sector and the for-profit sector need to put this goal at the top of their list of priorities and do nothing that will impede it. (I am proposing here the Internet in numerous instances will propagate more swiftly guided and wielded by those to whom profit is the motivator, but that they will need to curb their urge to restrict or co-opt by legal or technical means the growth of the overall network—that this is something that is in the long-term interest even of those who hope to profit.)" —**William B. Pickett**, Rose-Hulman Institute of Technology

"Java-enabled people. By this I mean devices implanted in our bodies that interact with other devices around us; in general, a networked world, in which most devices are connected to the Internet." —**Michael Wollowski**, Rose-Hulman Institute of Technology

"Presence. The ability to be elsewhere. Some forms of virtual reality—or at least distance-connectivity; not just the avatars and hybrids, but deeper, richer connections with old friends or new contacts—people-based not just machine-generated. Education is high priority. Not just learning, but also just-in-time knowledge. Ability to find new ideas and see them, which then extends to all kinds of applications in health/medicine, business, science, job performance." —**Gary Arlen**, Arlen Communications Inc.

> "I am anxious to see a change in copyright law for all concerned. My dream application would be something that creates access to all media, library-like, regardless of format. Everyone has access; everyone gets paid. It is in the public interest to move this way. I think open systems are a good thing. There should be alternatives."
>
> —Sam Punnett,
> FAD Research

"I am anxious for all of the security concerns related to the Internet to be completely eliminated. I would like to see the cost of high-speed access fall considerably. I would like to see the resolution of security concerns and cheap bandwidth; see more widespread use of personal servers and all of the advantages that that would offer to small businesses, households and institutions such as schools. The advent of cheap, ubiquitous, high-bandwidth wireless will magnify the impact of the Internet." —**Ezra Miller**, Ibex Consulting, Ottawa, Canada

"Communication via entrance to a virtual-reality-style metaverse." —**Ben Fineman**, Internet2

"I hope to see as soon as possible that everyone makes a video call as easy as picking up the phone now. I think the marriage of media, entertainment, and education delivered through the network on the device of your choice will further develop people (and therefore mankind) everywhere in the world, not just the developed countries. The now underdeveloped countries will skip generations of network and IT technology and quickly catch up (see the growth of wireless networks in Africa and India), giving them at least equal chances." —**Egon Verharen**, innovation manager, SURFnet (Dutch national education and research network)

"First, I would hope that the commercialization of the Net will taper off, though I am not so naive as to believe that this will actually happen. I would like to see the Internet being used more to affect positive change particularly in terms of sustained development and conflict resolution, particularly in the Middle East, and particularly in the framework of the Israeli–Palestinian conflict. I feel that the potential is there but it has not been harnessed as of yet." —**Michael Dahan**, Ben Gurion University of the Negev, Department of Comparative Media, Israel

"Greater reliance and adherence to standards allows more people easier and more efficient methods to publish and share information and experiences with each other. Commercialization of the Internet slows so that individuals and social concerns are able to express their needs and interests without being forced to 'consume' Internet resources as television and radio have increasingly done." —**Roger Seip**, EDS

"My dream is that the American population can be more educated information consumers, more discriminating about what they hear and believe, and will take personal control of their lives and choices with less control by centralized media, entertainment, commerce, or government." —**Dan Ness**, MetaFacts

"I am anxious to see a system that will help evolve our democracy. I have been very hopeful about the way the Internet has given a two-way form of communication for citizens with the government. However, we need to trust the information and infrastructure more. As far as my dream application—the universal scheduler to help truly make our lives easier. I also hope to see people set boundaries with technology so that there are times when we are 'off.'" —**Tiffany Shlain**, founder, the Webby Awards

"I'd like to see a greater voice for the unheard, more 'channels' devoted to alternative ideas, and a greater commitment to equitable distribution of the skills and infrastructure required to fully participate in the Internet-enabled world of the 21st century." —**Laura Breeden**, Education Development Center

"Ranking systems that enable people to find content considered useful by others like themselves will level communications and learning systems and enable faster, more solid societal development." —**William Stewart**, LivingInternet.com

"My dream would be to free musicians and music fans from the clutches of the Big Five record labels by creating an environment where musicians and fans can interact with each other in a mutually respectful way that pays musicians for their work and offers many listening and purchasing choices to fans. One great promise of the Internet is to afford musicians a means of distributing their own music without the censorship, artistic interference, and market stranglehold historically wielded by record companies. Hopefully, by 2014, many more musicians will be able to avoid being called, 'hey waiter!'" —**Peter W. Van Ness**, principal, Van Ness Group

"Cross-integration, worldwide, of library and other databases and reference tools, peer-organized (i.e., not corporate controlled, but publicly owned), auto-updating, and self-repairing. If I buy a book on Amazon or a music track on iTunes (or whatever), I want it to be added automatically to my personal library catalog, without me having to enter this information manually. I want this catalog to include everything that I own, to be saved automatically on the Net, to be acces-

> "A simple, cheap, and effective means to create an entire Web site from scratch without having to learn HTML, Java, Shockwave, XML, etc."
>
> —Graham Lovelace, Lovelacemedia Ltd., U.K.

sible only to myself and those I designate—and I don't want Microsoft to own the passport. I want everything in my personal library to be automatically hyperlinked to the rest of the world, so that I can jump immediately, and without configuring anything, from a passage in a printed book that I own to its electronic representation and on to the manuscript it references out there in Timbuktu. I was asked to dream, right?"
—**Albrecht Hofheinz**, University of Oslo

"I would like to see technology break down the biases and ignorance barriers people hide behind. The fruits of technology should make people smarter and morally better. Right now, technology sometimes contributes to stupidity, bigotry, political polarization, and predation." —**John Mahaffie**, cofounder, Leading Futurists LLC

ANONYMOUS COMMENTS

The following excellent contributions to this discussion about what people are "anxious to see happen" are from predictors who chose to remain anonymous.

"I am anxious to see new developments in technology that ease human suffering rather than increase individual and corporate profit."

"I would like to see increased education and opportunity in Third World countries thereby improving the security of everyone."

"I'd like to see the Internet deliver the most influential and advantageous information to the greatest number of people at the lowest price that would make the most difference in the state of the world. The Internet as hyperlever."

"Independent media, revolution in copyright and intellectual property laws, more independent applications, open-source software, integrated technology (i.e., cellular phone, music player, camera, e-book reader, Web browser, e-mail all in one powerful, portable device)."

"An end to junk e-mail and spam." —*This was an anonymous answer from more than a dozen survey participants*

"Secure communication. Note that this is a hope rather than a prediction."

"No screens, no keyboards, no stylus, no stinted speech. I want to be able to ask for something and get it."

"I'd like to see a market emerge for solid empirical evidence about a host of issues that people have to decide on every day. I'd like to see people becoming more skeptical consumers of evidence as well."

"I'd like to see a global awareness develop around poverty, the environment, and human rights abuse that unites people to develop solutions."

"Virtual reality."

"A single, portable, wireless device that allows access to all information sources (data, records, moving image, sound). The device and its use are inexpensive and easy."

"That RSS really becomes the third platform (joining Web and e-mail) on everyone's computer. Ninety percent of regular Internet users visit a handful of content-driven sites the majority of the time. It does not make sense that we need to seek out information when it can be sent to us directly (i.e., RSS) without clogging our inboxes."

"Electronic paper in small, flexible, and portable format, with a high enough resolution to display graphics and video, and with a wireless connection to the Internet. When that occurs, the distribution side of entire publishing industry will be turned upside down, much as the film processing industry is being undermined by digital photography now. Then, the full impact of digital technology on storytelling and reporting will really be felt."

> "Educating the Millennial and Cyber generations equally so that when they reach a college age they're all smart enough to go, and after college they can all get good jobs because the educational system throughout their lives kept pace with technology, and the teachers did as well. No kid should be left behind."
>
> —Anonymous respondent

"The easy capture of body data in emerging technologies for a true consumer-centric health care to emerge. See Eric Dishman's work at Intel."

"More at the edge—wireless sensor Nets. More portability/ubiquity."

> "Right now, my primary focus is on harnessing the power of computer and video games to enable new forms of teaching. We see strong signs that parents and teachers are ready to embrace such technologies in the classroom, while ironically the resistance is coming from within the games industry where people are frightened of the 'L word' [learning] and unwilling to risk their status as an entertainment medium to take on new roles."
>
> —Anonymous respondent

"I am anxious to see security solutions deployed to address spam, viruses, and worms. I am looking forward to the deployment of IP telephony and real-time communications, which will provide more sophisticated, less expensive, and more customizable solutions for consumers and business."

"No applications, just better use of resources to share what we have throughout the world so that the widening gulf between haves and have-nots does not continue to increase, with all the attendant fallout such as wars, hunger, and disease that come from this basic structural social problem."

"In education, Internet access is not enough; having the knowledge and wherewithal to use this tool effectively for education is imperative. Further, Internet access is becoming universal but the ability to use it productively to learn is still divided by socioeconomic groups."

"'Brain amplifier'—the integration of computers as a way to boost human intelligence."

"The video Internet: Video e-mail, telephone, conferencing, publishing, education, merchandising."

"Cheap, reliable Internet appliances to replace PCs; network security getting appreciably better, not worse; spam eradicated; sensible IPR balance between content owners and users."

"A centralized authentication database of all copyrights both privately held and those in the public domain, and then a decentralized competitive licensing structure that truly allows for technical creativity and competition in the marketplace, circulation of ideas, art, and information while at the same time compensating artists."

"Celestial jukebox—all media available all the time. Brewster Kahle's idea of universal access to all human knowledge."

"Telepresence, i.e., significant improvements in technology that enable humans to interact in real time. Images and sound help, but only the latter is effective in real time. A significant breakthrough related to video interactions would be important. I would like to see a much more balanced approach to intellectual property that enables and encourages creative reuse. I see this as a policy problem, not a technology problem. Technology favors the copyright holder, at the moment. I'd like to see automatic language translation play a role in enabling access to more worldwide content, particularly to news reporting."

"The expansion of niche audio networks, providing many new formats; that seems to be coming, although slowly. I would really like to see the major noncommercial media companies in the world—the government broadcasters and BBC, CBC, etc.—combine forces to create a much larger noncommercialized space on the Web. I definitely fear the commercialization of all online services. I would like to see a substantial expansion in the access to large databases of news and information for younger students (Lexis-Nexis, ProQuest, etc.) I think that the search capability of the Internet, which may be its most powerful attribute, is severely restricted by current level of access on the part of the general population (and younger students in particular). I would like to see a substantial expansion of high-speed access in rural areas."

> "Complete access to all human knowledge."
>
> —Anonymous respondent

Looking Back; Looking Forward

"Digital rights management and legal frameworks."

"If there's a dream, it's for more thorough search engines. That's probably still one of the most primitive exercises being conducted today in light of the computing power available. Search results will almost certainly be perfected in the 10 years. My dream application would be a device to measure health on a daily basis and send it to your doctor, providing an early warning of potentially worrisome issues."

"I am most anxious to see a 'grand unifier' device for communications and information use, a device that allows me the to choose the cheapest or most reliable or most secure network to place a call; a device that is connected to 'the Net' anywhere I go; a device that is useful for research, writing, communication, storage, entertainment, commerce; a device that connects me instantly to my family, screens my calls/messages, alerts me to needed changes in my schedule, allows me to control appliances in my house, etc. And, of course, it needs to allow me to track Cubs games wherever I am as they finally put together a season that ends in a World Series victory."

> "There must be a way, without violating civil liberties, of finding hackers, cyber terrorists and others who are criminal in nature. If the Internet can detect where the bad guys are, it can be a safer world and a safer Internet."
>
> —Anonymous respondent

"Better communications; less wiring; an OS that does not crash."

"Way back in the '80s, Apple devised a concept called the Knowledge Navigator. It featured a human-looking 'agent' that managed appointments, searched for information, provided reminders, etc. That's my dream application…something that automates tasks. You can do it now through macros, but I am still surprised how much pointing and clicking or typing I need to do to say flip through pages in a sequence, or to look for similar files."

"Mechanisms to set desired level of spam, more or less credible information, link statements to facts, verify content."

"(1) Wireless homes and workplace—the disappearance of the 'tangle'; (2) ubiquitous on-demand availability of the entire category of musical and video products."

> "I want my digital life accessible, wherever I happen to be, without lugging around a lot of machines and wires."
>
> —Anonymous respondent

"The most impactful application would be a browser capable of accurate and fast on-the-fly language translation of Web sites and data. The depth and breadth of the Internet would then truly be global—and world-changing."

"The biggest things holding back the wider application of the Internet are; (1) solid but practical form of end-user authentication to reduce fraud and eliminate spam, (2) 10-megabit-class, final-mile connectivity at home."

"I would hope to see 'big broadband' widely deployed with open architectures that allow for vigorous competition at the application and content layers. I fear that instead we will get an oligopoly with highly constrained access."

"A key challenge is the ability to integrate all the information, all the communication possibilities, into a sustainable lifestyle. How does the Net produce music for us when we want or need it rather than produce a constant unbearable noise? How do we avoid being paralyzed by the choices and opportunities? How can the world feel simple when we can see its deepest complexities 24/7?"

"I want to see widespread adoption of IPv6, for security enhancement purposes. I believe widespread Internet telephony (VoIP) will have tremendous effects, including 'all-you-can-use-for-one-price' telephony, which in turn will destroy most current major telephone companies."

"(1) Rapid expansion of public health and medical applications. Telemedicine could help provide access to specialists. (2) Rapid and extensive deployment of fiber connections to the home and office. (3) New user interfaces and display technologies. Voice- and identity-recognition applications. (4) The 'Ask Jeeves' search-engine concept on super steroids. The ability to search, retrieve, and organize information from multiple, multiple formats (print, image, audio, video, database) using intelligent filters for relevance, importance, reliability, and validity. (5) Artificial intelligences in appliances, vehicles, computer software."

> "Using the Internet to allow people of the world to interact with each other instead of being disembodied stereotypes. Unfortunately, corporate and government control will never allow free access."
>
> —Anonymous respondent

"More news outlets lead to more truth-telling, especially in government and politics. I would hope that the Internet would allow for health care access to all Americans, and that this technology could help us solve health problems quicker—like cancer and AIDS."

"I would like to see a more global equality with regard to connectivity. Still, large parts of the world are not connected, or at ridiculously low dial-up speeds and/or horrendous prices. We cannot talk about a global village until we get closer to that."

"Organization of personal information. Reducing information overload and IT-related stress. Improvements in usability and comprehensibility of ICT. Transformation of health care. Improvement in quality of life for the elderly. Penetration of ICT into the developing world."

"Free and unrestricted access for all humans in all countries."

"Return to the principals of the computer inquires of an open, physical communications network open to all. Provision of true broadband at realistic prices. Recognition of and restraint of significant market power."

"My dreams are: universal access on a par with water and electricity as a public service; a variety of multimodal interfaces, including speech/sound, that will enable non- or semi-literate individuals to communicate in more social contexts; developing effective and acceptable boundaries between home and work life."

"I would like to see it become a bit easier for ordinary, non-technically empowered folks to start online organizations, à la open source. We see the beginnings of this in the ability to start your own mailing list, etc., but it would be nice to see more powerful capacities available."

"More citizen engagement in public policy debates and formulation; more precise searching capabilities. Path-breaking: for Internet devices to be greatly enhanced so that connections are consistently available, devices are truly easy to operate, and offered at low costs. If that happens, all manner of content follows."

"I would like to see greater Wi-Fi access and faster Wi-Fi. I would like to see more and better networking options, especially at home. I would like to see an easier way to link up television, computer, DVD, cell phone, Internet all together."

"Device proliferation with always-on connection to the Internet. Specialized devices that serve any and every human need."

"Easy interfaces for handheld devices that create pervasive access to Internet."

"Proper ad models for online publications. User acceptance that content costs money. Functional, spam-free e-mail systems. E-voting. Reduced network access costs, and firm anti-oligarchic regulations designed to prevent another regulatory mess, à la cable TV. Better integration of open standards like MP3. Microsoft finally matched in its monopoly. AOL sidelined by superior technology."

"Driving down of the cost of health care and widening of its availability. Adoption of the Net as the conveyor of written material in a format that resembles the page. Convergence of communications capabilities into a single device that enables wireless audio-, video-, data-, and voice-transmission and reception."

"I would be anxious to see the Internet facilitate voting in elections and energizing the electorate to be more participatory. Unfortunately, voter turnout, or the lack thereof, is not a technology problem."

"I want the social force and interface power of Google to outstrip all firewalls and proprietary systems, to make more works accessible to all. Yet I don't want that much power concentrated in such a central location like Google, so I really look to the semantic Web and XML to provide metadata that will enable many kinds of powerful searching outside of Google, or perhaps by those using Google tools to mine data for their own ends. I don't yet fear Google, but I could. I'm also antsy for text-based tools like RSS and Atom, which are so powerful, to work for interactive media artifacts, as well: audio, video, Flash. Right now the power of text-based interfaces is actually discouraging higher-bandwidth forms of communication. Usability is getting far too rule-based, prescriptive, and entrenched. The move toward CSS-template-driven content-management systems was welcome and empowering, but we've lost as much as we've gained, in creativity, in bandwidth-sucking bells and whistles, and in interactivity."

"There is so much information and knowledge out there that could help developing countries…Time zones don't matter, no phone bills to pay, you can even send video mail if you want to see, as well as talk…I'd like information-sharing amongst scientists to continue, especially in areas like weather and medicine…By sharing information about each other all around the world, people from all cultures can better-understand each other, and we learn to live and let live. Basically, I think the Internet can help mankind to make our world a better place."

PART 18

REFLECTIONS

Conflicting desires can be seen in these expert opinions, mirroring basic concerns that were discussed at length in the predictions made by Internet stakeholders in the early 1990s.

First is the conflict between the all-too-human desires for total security and complete privacy. Everyone wishes for both, but it is impossible to have both and make things work online. For instance, there is no way to have secure online voting without asking voters to reveal their identities as they vote, thus removing their anonymity. At the same time, there cannot be total privacy and anonymity online without allowing criminals and terrorists to operate in secret. If there is not a reasonable level of security, Internet users cannot trust that companies on the Internet will be able to handle the most personal kinds of transactions. Security does not have to be perfectly air-tight, argue some experts, but it has to be effective enough to allow people reasonable confidence in Internet-derived information and transactions.

Second is the conflict between the yearning for access to all information everywhere and the desire to simplify life and avoid being inundated with information. Many of the same people who express a desire for total, instantaneous, easy-to-use access to all of the information on the planet also complain that the avalanche of information is growing worse all the time, complicating their lives, causing stress, and even changing the dynamics of work, family, and leisure time.

Finally, our experience in collecting the original material for the 1990–1995 predictions database and then going through the results of this survey remind us that some universal themes about the impact of technology and society are evident here. Respondents' comments reflected a number of the same concerns expressed about previous technologies throughout history, among them were:

1. Technological change is inevitable, and it will result in both beneficial and harmful outcomes. Those surveyed see the impact of the Internet as multidirectional and complex, as did predictors at the dawn of all other communications technologies.
2. A technology is never totally isolated in its influence as a change agent. Many social trends commonly associated with the coming of the Internet are the result of changes spurred by multiple forces; some already were in motion as the Internet came into common use. We must not fall into the trap of technological determinism—the Internet should not be fully credited nor should it take all of the blame.
3. Entrenched interests prefer the status quo and often work to block or delay innovations introduced by new technologies such as the Internet. Respondents see this happening in copyright clashes, education, health care, and other areas.
4. The business of projecting the future impact of a technology can be difficult and full of inconsistencies. Respondents' answers display a conflict between their hopes for the Internet's positive potential and their reality-based opinions of what can really be accomplished in the next 10 years. Many were skeptical about advances outside their areas of expertise and were enthusiastic about those in their areas of specialization. Opinions diverge on many issues.

It is appropriate to close this report with a quote from one expert who wrote, "I never would have expected that such a high percentage of people would be utterly dependent upon the Internet for such a large proportion of their daily communication activities. If you took it away, we would be shell-shocked. But 10 years ago, we didn't even have it!"

METHODOLOGY

This survey, sponsored by the Pew Internet & American Life Project and conducted by Princeton Survey Research Associates International, obtained online interviews with a nonrandom sample of 1,286 Internet users. The interviews were conducted online, via SPSS, in two waves: Wave 1 took place from September 20 to October 18, 2004, and Wave 2 took place from October 19 to November 1, 2004. Details on the design, execution, and analysis of the survey are discussed next.

SAMPLE DESIGN AND CONTACT PROCEDURES

Across both waves of the project, e-mail invitations to participate in the survey were sent to slightly less than 1,000 Internet users (367 of these were sent after the completion of Wave 1 of the project). The initial list included as many members as possible from the "200 Internet Figures" identified in the Elon University/Pew Internet & American Life Predictions Database project (http://www.elon.edu/predictions/200briefbios.aspx). Overall, approximately 7% of the e-mail addresses proved invalid, for a working rate of 93%. The e-mail invitations provided a direct link to the survey, and contained the following language:

> Dear [name here]:
>
> The Pew Internet & American Life Project is surveying experts about the future of the Internet and we would very much like to include your views in our research.
>
> The idea for this project grew out of work we did with Elon University to develop a database of over 4,000 predictions about the impact of the Internet made by experts during the period between 1990 and 1995. Now we are conducting a Web-based survey about the impact the Internet might have in the next decade. We are canvassing many of the people whose predictions are included in the original 1990–1995 database—and we are soliciting predictions from other experts who have established themselves in recent years as thoughtful analysts.
>
> We hope you'll take 10–15 minutes to fill out our survey, which you will find at http://surveys.spss-sb.com/spssmr/survey/surveyentry.aspx?project=p3280003. The survey asks you to assess several predictions about the future impact of the Internet and to contribute your own thoughts about what will happen in the next 10 years.
>
> This is a confidential survey. However, we encourage you to take credit for your thoughts. When you start the survey, please use this personal identification number (PIN): [PIN]
>
> The Pew Internet & American Life Project will issue a report based on this survey during autumn; we hope the results will be useful to policy makers, scholars, and those in the information technology industry. Our goal is to include material from this new survey in the predictions database. (While we have not publicly talked about that effort yet, you can browse through the existing material at http://www.elon.edu/predictions.) Be assured that we will not use your name or e-mail address for any purpose other than this research project, and will not share your information with outside solicitors.

> *We're sure we have not identified all experts whose views would be helpful to this research, so I would invite you to send an invitation to participate in this survey to any friends or colleagues whose insights you would be interested in learning. Please ask them to use PIN 700 when taking the survey.*
>
> *I hope you enjoy taking the survey and sharing your views about the future of the Internet. If you have any questions, please feel free to contact me at lrainie@pewinternet.org.*
>
> *Thank you,*
> *Lee Rainie, Director, Pew Internet & American Life Project*

As the aforementioned text indicates, Pew Internet encouraged the initial sample of experts to forward the e-mail invitation to any colleagues whose thoughts on the future of the Internet they would consider useful and important. This created an additional snowball sample of Internet experts, whose ideas are also included in the final data.

COMPLETION RATE

Based on figures supplied by SPSS, PSRAI has calculated the completion rate for the Experts Survey as seen in Table 1.

In Table 1, total hits (1,892) indicate the number of times the survey link was accessed between September 20 and November 1, 2004, or roughly the number of potential respondents who reached

TABLE 1. Overall survey completion rate.

	Number	*Rate*
Total Hits	1,892	
Total Completes	1,286	68.0%
Final Completion Rate		*68.0%*

the survey's title page during the field period. The survey title page gave the following brief description of the survey and its sponsors, along with instructions for how to complete the survey:

> **Forecasting the Internet**
>
> *Welcome to the Pew Internet & American Life Project survey of technology experts and social analysts about the future of the Internet. This survey asks you to assess some predictions and contribute your own thoughts about the impact of the Internet in the next 10 years.*
>
> *This survey has grown out of as yet unpublished research by the Project and Elon University to study predictions made between 1990 and 1995 about the evolution of the Internet. The "Imagining the Internet" database of those predictions is available at http://www.elon.edu/predictions. We plan to update the database to include responses from this survey, as well as your unfiltered answers.*
>
> *The Project's goal is to see where experts agree and disagree about the potential social impact of the Internet. We hope the findings will illuminate issues for policy makers, spark debate and further research among scholars, and encourage those who build technology to ponder the societal effects of their creations.*
>
> *This is a confidential survey. However, we encourage you to take credit for your thoughts. Please feel free to put your name in any space that allows for written answers. We will only credit to you the individual statements to which you add your name in the answer block. If your name is not there, your comments will be attributed to an anonymous voice when they are added to the Pew Predictions Database.*
>
> *We plan to publish the results of this survey in a report that will be issued this autumn.*
>
> *S1. If you received an e-mail invitation from Pew Internet with an individual PIN for taking this survey, please enter it here.*
>
> *Those who were invited to participate by a friend or colleague should use guest PIN 700. If you did not receive either an individual or guest PIN, please enter 999 and proceed.*

Total completes (1,286) indicate the number of respondents who completed the survey through at least Question 6. The final completion rate for the survey is computed as the number of completes (1,286)/the number of hits (1,892), or 68.0%.

QUESTIONNAIRE DEVELOPMENT

The questionnaire was developed by PSRAI in collaboration with staff of the Pew Internet & American Life Project and their partners at Elon University.

Appendix I

Future Scenarios Inspired by Early 1990s "Awe Stage" Projections

Information gathered to build the Elon University/Pew Internet Early 1990s Predictions Database was a partial inspiration for the formulation of the questions for the initial Future of the Internet survey. The 2004 survey questions were purposely constructed in a many-layered manner to spur discussion. They are rooted in a series of previous predictive statements. They do not represent the beliefs or research conclusions of any of the researchers. This section offers some background—a back story of sorts—tied to the 2004 questions to help put things in context.

Prediction on Social Networks

> **Scenario:** *By 2014, use of the Internet will increase the size of people's social networks far beyond what has traditionally been the case. This will enhance trust in society, as people have a wider range of sources from which to discover and verify information about job opportunities, personal services, common interests, and products.*

In the data in the Early 1990s Predictions Database, the general drift of the commentary skews in agreement with this prediction on social networks. However, some predictors saw problems with the potential for a digital divide, overt commercialism, and/or the evils associated with life in a high-speed world to spoil this dream. Here are a few selections:

- In his 1995 book, *Democracy and Technology*, Richard Sclove writes, "One function of democratic community is to provide a social foundation for self-governance and individual political empowerment. This suggests that community boundaries ought normally to remain roughly contiguous with the territorial boundaries defining formal political accountability and agency. Yet the criterion of local self-governance is breached if involvement in spatially dispersed social networks grows to subvert a collective capacity to govern the locales people physically inhabit. And the criterion of egalitarian empowerment is breached if coveys of technorich cronies are empowered to telelobby senators, while technopoor neighbors are excluded from the circuit" (p. 80).

- In a 1995 online essay, Justin Hall makes the following statement, "I encourage all my friends in the commercial sector to be generous, and trust that their product is worth talking about. Leave the channels open for people to do so. Otherwise, the Internet will accelerate the self-loathing and dissatisfaction that comes with advertising's endless call for immediate gratification. Identified by and targeted for our product consumption, we will find ourselves receiving more personalized mail from products than from people. They will know us, and they will manipulate us. We will end up hating the Internet, and ourselves" (Hall, 1995b).

- In a 1995 article for *The Nation*, adapted from his book, *Rebels Against the Future: The Luddites and Their War on the Industrial Revolution*, Kirkpatrick Sale urges people to step back and question technology: "This transformation is, without anyone being prepared for it, overwhelming the communities and institutions and customs that once were the familiar stanchions of our lives...No wonder there are some people who are 'Just Saying No.' They have a great variety of stances and tactics, but the technophobes and technoresisters out there are increasingly coming together under the banner that dates to those attackers of technology of two centuries ago, the Luddites...These would include those several million people in all the industrial nations whose jobs have simply been automated out from under them or have been sent overseas as part of the multinationals' global network, itself built on high-tech communications...They may include, too, quite a number of those whose experience with high technology in the home or office has left them confused or demeaned, or frustrated by machines too complex to understand, much less to repair, or assaulted and angered by systems that deftly invade their privacy, or deny them credit, or turn them into ciphers. Wherever...neo-Luddites may be found, they are attempting to bear witness to the secret little truth that lies at the heart of modern experience: Whatever its presumed benefits, of speed, or ease, or power, or wealth, industrial technology comes at a price, and in the contemporary world that price is ever rising and ever threatening" (p. 785).

PREDICTION ON ATTACKS ON NETWORK INFRASTRUCTURE

> SCENARIO: *At least one devastating attack will occur in the next 10 years on the networked information infrastructure or the country's power grid.*

In the data in the Predictions Database (1990–1995), the commentary skews in agreement with the prediction that malicious attacks could be launched on the system. Individuals who discounted the likelihood of attacks tended to be people who spoke out against government regulation (through key-escrow encryption, etc.) that could take away civil liberties. Here are a few examples:

- In a 1995 article for *The New York Times*, John Markoff talks with Eric Schmidt, chief technology officer for Sun Microsystems, regarding computer viruses. Markoff writes: "'I think [viruses] will be an extraordinarily serious problem over the next few years,' said Eric Schmidt, chief technical officer at Sun Microsystems Inc. 'If you believe the theory that nearly all personal computers will be on corporate networks or online services in the next 2 or 3 years, then this is a problem that could touch all PC users worldwide...There are criminals in the world, and some of them are programmers. With computer networks, they have an amplifying effect that they've never had before. If I were a criminal with a gun, I might attack one person. But with a computer network, I can attack a million people at a time. It's like an atomic bomb'" (Lewis, 1995, p. 1).

- In their 1994 book, *Firewalls and Internet Security: Repelling the Wily Hacker*, William Cheswick and Steven Bellovin write: "The advent of mobile computing will also stress traditional security architectures. We see this today, to some extent, with the need to pass X11 through the firewall. It will be more important in the future. How does one create a firewall that can protect a portable computer, one that talks to its home network via a public IP network? Certainly, all communication can be encrypted, but how is the portable machine itself to be protected from network-based attacks? What services must it offer, in order to function as a mobile host? What about interactions with local facilities, such as printers or disk space? The face of the network security problem will certainly change over the years. But we're certain of one thing: it won't go away" (pp. 236–237).

- In a 1995 article for *Wired* magazine, Peter Schwartz, cofounder and president of Global Business Network and author of *The Art of the Long View*, interviews Andrew Marshall, a national security researcher/consultant whose work included stints at the RAND think tank in 1949 and 22 years at the Pentagon, under six presidents. Schwartz quotes Marshall saying: "There may well be an increase in guerrilla warfare because new technologies may increase our vulnerability to it. We are living in the equivalent of the early 1920s, when tanks, airplanes, and later radar and radio were new, and people weren't sure what they were or how to use them. We have only preliminary ideas about how today's technology is going to change warfare. But it will. In the old world, if I wanted to attack something physical, there was one way to get there. You could put guards and guns around it, you could protect it. But a database—or a control system—usually has multiple pathways, unpredictable routes to it, and seems intrinsically impossible to protect. That's why most efforts at computer security have been defeated" (p. 138).

PREDICTION ON DIGITAL NETWORKS

> **SCENARIO:** *In 2014 it will still be the case that the vast majority of Internet users will easily be able to copy and distribute digital products freely through anonymous peer-to-peer networks.*

In the data in the Early 1990s Predictions Database, the commentary skews in agreement with the prediction on freedom to copy digital products. Kelly, Dyson, and Barlow were the big names speaking out on this topic, but there were others. Here are a few:

- In an article she wrote in 1995 for *Wired* magazine, Esther Dyson comments on intellectual property rights in the future on the Internet: "In the new communities of the Net, the intrinsic value of content generally will remain high, but most individual items will have a short commercial half-life. Creators will have to fight to attract attention and get paid. Creativity will proliferate, but quality will be scarce and hard to recognize. The problem for providers of intellectual property in the future is this: Although under law they will be able to control the pricing of their own products, they will operate in an increasingly competitive marketplace where much of the intellectual property is distributed free and suppliers explode in number" (p. 137).
- In a 1994 article he wrote for the newspaper, *The Guardian of London*, *Wired* magazine editor Kevin Kelly says, "Let the copies breed. Whatever it is that we are constructing by connecting everything to everything, we know the big thing will copy effortlessly. The I-way is a gigantic copy machine. It is

a law of the digital realm: Anything digital will be copied, and anything copied once will fill the universe. Further, every effort to restrict copying is doomed to failure...Controlling copies is futile. This presents a problem for all holders of intellectual property who adhere to the notion of copyright—such as Hollywood moguls and authors. Copyright law as we know it will be dead in 50 years. A legal system that shifts the focus from the 'copy' to the 'use' must take its place, letting copies proliferate, and tracking only how and when an item is used. Copy this article, please!" (Kelly, 1994a, p. S2).

- In a 1993 article for *Internet World*, Mike Godwin, chief counsel for the Electronic Frontier Foundation, outlines issues in regard to law and the Internet. Godwin writes: "The law of intellectual property, which includes the law of copyright, will have to adapt to a world in which advances in technology increasingly undercut one's ability to enforce intellectual property rights. Only now has it become clear the extent to which copyright law has depended on the 'bottleneck' created by the costs of printing (and, later, of photocopying). As of today, it remains far easier simply to buy the magazine containing a short story than it is to photocopy that story, or to have a copy printed by a printshop at your own expense. But...the costs of reproducing all sorts of intellectual property are falling rapidly...It has been argued that the current copyright regime could be replaced by one based on usage fees, but that suggestion overlooks a couple of important obstacles. First of all, once someone acquires information from an online publisher, there's little disincentive to spread that information around. (Why should you call up Nexis, for example, when I did a similar search last week and can forward my search results to you in e-mail?) The second problem is that both the Internet and the proposed infrastructural schemes that could replace it are highly decentralized. This decentralization of the Net makes billing for and tracking use of intellectual property very difficult...I have long believed that, when a law's

requirements are so unrealistic that they are routinely broken by otherwise law-abiding citizens, it's a sign that the law needs to be changed" (pp. 52–54).
- In 1994 John Perry Barlow wrote an article for *Wired* that he described as, "a framework for rethinking patents and copyrights in the Digital Age." In this section, "Information is a Life Form," Barlow looks at how information changes and how difficult it is to copyright this evolving form: "Our system of copyright makes no accommodation whatever for expressions which don't become fixed at some point, nor for cultural expressions which lack a specific author or inventor. Jazz improvisations, stand-up comedy routines, mime performances, developing monologues, and unrecorded broadcast transmissions all lack the Constitutional requirement of fixation as a 'writing.' Without being fixed by a point of publication, the liquid works of the future will all look more like these continuously adapting and changing forms, and will therefore exist beyond the reach of copyright...Soon, most information will be generated collaboratively by the cyber-tribal hunter-gatherers of cyberspace. Our arrogant dismissal of the rights of 'primitives' will soon return to haunt us" (pp. 90, 126).
- In a 1994 article for *The Toronto Sun*, Scott Magnish talks with Lance Hoffman about law on the Internet. Magnish writes: "The concept of 'copyrighting' could be lost on the information highway as the world moves closer to the free flow of information, U.S. experts said...'Does copyright have a chance?' Hoffman asked rhetorically. 'I'm increasingly leery of the pressure. The economic pressures and even some of the social pressures are such that it may not. Maybe the whole nature of intellectual property has to be reexamined" (p. 51).
- In a 1995 report from a joint hearing of the House and Senate Judiciary committees Courts and Intellectual Property subcommittees, testimony from Sen. Patrick Leahy (D-Vermont) includes the following statement: "We must update our copyright laws to protect the intellectual property rights of creative

works available online. The future growth of computer networks, like the Internet, and of digital, electronic communications require it. Otherwise, owners of intellectual property will be unwilling to put their material online. If there is no content worth reading online, the growth of this medium will be stifled, and public accessibility will be retarded" (*NII Copyright Protection Act*, 1995).

PREDICTION ON CIVIC ENGAGEMENT

> **SCENARIO:** *Civic involvement will increase substantially in the next 10 years, thanks to ever-growing use of the Internet. That would include membership in groups of all kinds, including professional, social, sports, political, and religious organizations—and perhaps even bowling leagues.*

In the data in the Early 1990s Predictions Database, the general drift of the commentary skews in agreement with the prediction, but there were plenty of people concerned that online involvement would cause problems, including a reduction in important human-to-human interaction. The predictions were grouped under the subtopic of "virtual communities." Here are a few:

- In a 1993 article he wrote for *Newsweek* magazine, Howard Rheingold says, "If we don't lose the freedom to speak as we choose, and if the price of access doesn't restrict virtual communities to the wealthy, we have the opportunity to build a grassroots electronic democracy. But first we have to understand the nature of the medium, its pitfalls, as well as its benefits. Virtual communities are not utopias...There are dark sides, just as every technology cast cultural shadows. Electronic bulletin-board systems can bring people together, but the computer screen can be a way of controlling relationships, keeping people at a distance. Words on a screen help people communicate without the usual barriers of prejudice based on appearance. That same distancing of real-life identity and online persona can lead to cybercads and charlatans who use the medium to swindle others. People are cruel and rude to each other in real communities—and human nature doesn't change because the community is mediated by a computer screen. Computer-mediated communications are particularly susceptible to deception...Every new communication technology—including the telephone—brings people together in new ways and distances them in others. If we are to make good decisions as a society about a powerful new communication medium, we must not fail to look at the human element."

- In a 1995 *Pittsburgh Post-Gazette* article about Carnegie Mellon University's HomeNet Project—a 3-year study of how 50 families were using the Internet—Steve Creedy quotes Robert Kraut, a professor of social psychology and human-computer interaction who was involved in the study: "Will the Internet expand people's parochialism by leading them to a wider range of people with the same interests, or will it encourage them to expand their interests to new areas? There are hints of both, but the jury's still out," according to Kraut. "It makes it more efficient to be parochial, but at the same time it gets you to come across people and interests that you wouldn't have simply by being in your small location with your previous identity," he says. "We're seeing both things happening, and we don't know which is going to be dominant" (p. C6).
- In a 1994 article for *Time Magazine*, Philip Elmer-Dewitt writes about conflicting views about the development and uses of the Internet: "Now, just when it seems almost ready for prime time, the Net is being buffeted by forces that threaten to destroy the very qualities that fueled its growth. It's being pulled from all sides: by commercial interests eager to make money on it, by veteran users who want to protect it, by pornographers who want to exploit its freedoms, by parents and teachers who want to make it a safe and useful place for kids...The danger, if this trend continues, is that people will withdraw within their walled communities and never again

venture into the Internet's public spaces. It's a process similar to the one that created the suburbs and replaced the great cities with shopping malls and urban sprawl. The magic of the Net is that it thrusts people together in a strange, new world, one in which they get to rub virtual shoulders with characters they might otherwise never meet. The challenge for the citizens of cyberspace—as the battles to control the Internet are joined and waged—will be to carve out safe, pleasant places to work, play and raise their kids without losing touch with the freewheeling, untamable soul that attracted them to the Net in the first place" (pp. 50–51, 56).

- In his 1995 book, *Democracy and Technology*, Richard Sclove writes: "Contemporary technological reporting is rife with notions of electronic communities in which people interact across regions or entire continents. Could such 'virtual communities' eventually replace geographically localized social relations? There are reasons to suspect that, as the foundation for a democratic society, virtual communities will remain seriously deficient. If the prospect of a telecommunity replacing spatially localized community ought to evoke skepticism or opposition, one can nevertheless remain open to the possibility of democratically managing the evolution of telecommunications systems in ways that instead supplement more traditional forms of democratic community. Caution is in order. However, the benefits of telecommunities can potentially include combatting local parochialism; helping to establish individual memberships in a diverse range of communities, associations, and social movements; empowering isolated or marginalized groups; and facilitating transcommunity and intersocietal understanding, coordination, and accountability. Systems designed to support such uses—especially without subverting local community—are unlikely to emerge without concerted democratic struggle" (p. 79).

- Kimberly Rose made the following statement in a research presentation at INET '95, the Internet Society's 1995 International

Networking Conference. She was a researcher with Apple Computer's Advanced Technology Group. She also worked with a consortium of schools in Southern California to develop collaborative dynamic curricula utilizing a wide-area telecommunications network. Rose remarks: "The potential the Web offers to build virtual communities is tremendous. Large and complex problems which concern us are now not only up to individuals to solve. By means of global networking on the Internet special interest groups and clubs are being formed. These groups can break down large issues into smaller ones and collaborate to solve problems."

- Christopher Scheer (1995) wrote the following in an essay for *The Nation*: "Take the future world of right-wing visionary George Gilder—please. Listening to Gilder, one might get the impression that the only thing keeping us from being happy is all these other people. If we could only live 'virtually,' we'd be safe from all the bad stuff out there and stimulated by all the good. His future is sort of an intellectual's version of the survivalist dream: Leave the now-irrelevant cities, hole up in your crime-free Utah faux ranch with your wall-size interactive TV and call up the world of high culture in Sensurround sound…I'm the goddamn wannabe Luddite who wonders what America will look like if every rich person has a sprawling compound in some gloriously beautiful—and ecologically fragile—state like Utah while cities are abandoned to the poor. And yet, I'm actually living the 'electronic cottage' dream of the Gilders, Gingriches, and Tofflers. I'm turning on, logging in and crashing out here in my own little nest. I'm a 'prosumer' in the infoweb, absorbing great gobs of data and disgorging a little of my own every day…I'm human, a social animal. I'm not a god, I'm a hairless chimp with a messianic complex and a mouse. I need human contact and simple pleasures. I need to eat, poop, and see people smile. I need some sun, some rain and the pleasures of someday holding the tiny paw of my own child. But instead of returning to the basics, I, like many of us, am spending more and more of my time with my face bathed in monitorglow, getting my fix of digital junk. Won't someone please unwire me before it's too late" (p. 634).

PREDICTION ON EMBEDDED NETWORKS

> **SCENARIO:** *As computing devices become embedded in everything from clothes to appliances to cars to phones, these networked devices will allow greater surveillance by governments and businesses. By 2014, there will be increasing numbers of arrests based on this kind of surveillance by democratic governments, as well as by authoritarian regimes.*

In the data in the Early 1990s Predictions Database, the commentary skews in agreement with the prediction on the rise in tracking by governments and business, and many activists were deeply concerned. In the database, the keyword "surveillance" conjures up dozens of hits. Among them are:

- In a 1995 article for *Wired* magazine, John Whalen does a bit of surveillance at the American Society for Industrial Security's annual convention, and quotes Roy Want, an inventor of 'active badges' and a scientist at Xerox PARC. Whalen writes: "[Roy] Want hails from England, the former empire that gave the world Jeremy Bentham, philosopher of utilitarianism and author of *Panopticon, or 'The Inspection House.'* Published in 1791, Bentham's treatise described a polygonal prison workhouse that placed the penal/industrial overseers in a central tower with glass-walled cells radiating outward. Mirrors placed around the central tower allowed the guards to peer into each cell while remaining invisible to the prisoners…More than 200 years later, Want, a computer engineer, has essentially reinvented the Panopticon. More accurately, his brainchild, known as the 'Active Badge,' would have made Bentham proud…Clipped to a shirt pocket or belt and powered by a lithium battery, the black box emits an infrared signal—just like a TV remote—every 15 seconds. Throughout the computer lab at the PARC, infrared detectors are Velcro-mounted to the ceiling and networked into a Sun workstation...While privacy tribunes see active badges as an ominous new development in the brave new workplace, Want and his colleagues see them as 'a double-edged sword,' with the potential for both benign and malignant uses...Want sees the tabs getting thinner and lighter. Each of us would have dozens scattered around the office, in the car, and at home. Detector cells will start appearing in public places or the home, he says. 'The device will tell you where you are, wherever you are.' Of course, it might also tell them where you are...'There are always these trade-offs between what's useful and what could be done to us,' says Want from the belly of the kinder, gentler Panopticon. 'The benefits to be had are so great; we just have to be sure that the people who are in control respect our privacy'" (pp. 84–85).

- In a 1994 article for *Computer-Mediated Communication* magazine, Stephen Doheny-Farina, a professor of technical communication at Clarkson University, writes: "Active badges should scare the daylights out of anyone. When it comes to connectivity, the employer must justify the surveillance. Everyone must assume that only extraordinary conditions merit surveillance. The requisite argument must not be, 'Why do you not want to wear the badge?' The requisite argument must be 'Why do you want me to wear it?' We must demand that the burden of proof is on the watcher, not the watched" (p. 18).

- In his 1994 book, *City of Bits*, MIT computer scientist William J. Mitchell writes: "Life in cyberspace generates electronic trails as inevitably as soft ground retains footprints; that, in itself, is not the worrisome thing. But where will digital information about your contacts and activities reside? Who will have access to it and under what circumstances? Will information of different kinds be kept separately, or will there be ways to assemble it electronically to create close and detailed pictures of your life? These are the questions that we will face with increasing urgency as we shift more and more of our daily activities into the digital, electronic sphere. Contention about the limits of privacy and surveillance is not new, but the terms and stakes of the central questions are rapidly being redefined. Isolated hermits can keep to themselves and don't have to keep up appearances, but city dwellers have always had to accept that they will see and be seen. In return for the benefits of urban life, they tolerate some level of visibility and some possibility of surveillance—some erosion of their privacy. Architecture, laws, and customs maintain and represent whatever balance has been struck. As we construct and inhabit cyberspace communities, we will have to make and maintain similar bargains—though they will be embodied in software structures and electronic access controls rather than in architectural arrangements. And we had better get them right; since electronic data collection and digital collation techniques are so much more powerful than any that could be deployed in the past, they provide the means to create the ultimate Foucaultian dystopia" (pp. 158–159).

- In his 1992 book, *Privacy for Sale*, Jeffrey Rothfeder writes: "In time, high-tech snooping and databanking could make earlier-generation activities seem naively old-fashioned, as innocent as child's play. When that occurs, our failure to legislate controls over surveillance equipment as they evolved—already a problem today—could overwhelm us, as could our failure to prescribe adequate civil and criminal penalties for abuses of individual privacy committed by government agencies and U.S. corporations" (p. 204).

- Jim Warren made the following statement in reaction to the fast-track passage of H.R. 4922 the "Communications Assistance for Law Enforcement Act" (called by some the "digital telephony bill" and labeled by its opponents as the "FBI's wire-tap bill") which provided rules for the "interception of

Future Scenarios 319

digital and other communications," in 1994. The law directed that all telecommunications companies make their networks tappable within in the next 4 years. The intent of the legislation, passed by the Senate and signed by President Clinton, was to aid law enforcement, but it included the phrase "and other lawful authorization," raising privacy questions. "How many tens of thousands of...[officials] will have authorized access to this pervasive surveillance power? How many thousands of political appointees control those agencies—and are controlled by incumbent politicians? How many hick sheriffs or local party bosses or nosy night staff are likely to make unauthorized use of this Congressionally mandated snoop-n-peep technology against boy and girlfriends, family members, personal enemies, business competitors, and—most dangerously—political opponents" (Lawrence, 1994).

PREDICTION ON FORMAL EDUCATION

> SCENARIO: *Enabled by information technologies, the pace of learning in the next decade will increasingly be set by student choices. In 10 years, most students will spend at least part of their "school days" in virtual classes, grouped online with others who share their interests, mastery, and skills.*

In the data in the Early 1990s Predictions Database, the commentary was enthusiastic about positive changes, as people projected the Internet to be a tremendous step forward in education. But they also warned that the new tool had to be seen in a different light, rather than just being swallowed up by the old system. Nearly 200 of the 4,200 predictions in the database fall under the education/schools category. Here are a few:

- In a 1993 *Wired* magazine interview with Connie Guglielmo about his election to the Colorado State Board of Education, Ed Lyell discusses his vision of the future of computers and education. Guglielmo writes: "In Ed Lyell's hopeful vision of the near future, by age 18 everyone in this country is literate, semi-skilled and as comfortable using computers and telecommunications technology as they are using pencils. The main thing clouding that vision is the current educational system. Children are born learning machines, says the Denver resident. '...But if you had a school out there today to teach children to walk, one-third of the population would not be walking...It would be easier to get the Pope to become a Buddhist than to get the schools to change.' Lyell has plans for an educational system where students are treated as individuals with differing interests and learning skills. He hopes to build interactive learning devices that students can peruse at their own pace and that present information in a variety of ways. These computer-based learning systems are part of a concept he calls 'Just-in-Time Learning.' 'It's analogous to just-in-time manufacturing, which holds that efficiency comes when things happen just at the right time, when you have all the proper resources in place,' says Lyell. 'In the case of education, it means a student is able to log onto a computer to learn about whatever he or she is interested in learning about at that particular point in time...I think we should have learning centers, neighborhood electronic cottages,' Lyell says."

- Kimberly Rose made the following statement in a research presentation at INET '95, the Internet Society's 1995 International Networking Conference. Rose worked with a consortium of schools in Southern California to develop collaborative dynamic curricula using a wide-area telecommunications network: "We must be careful not to look to this technology with hopes that it will be the next band aid for education. Installing computers, software, networking hardware, telephone lines, and cabling in our schools will not change the way our children think unless we use these tools in new ways which take advantage of the possibilities the new tools have to offer. It is more likely that these new technologies will be used in ways

which just mimic the old media and, therefore, not gain us any new insights into creating better learning environments."
- In a 1993 article he wrote for *Wired* magazine, Seymour Papert remarks: "In the past, education adapted the mind to a very restricted set of available media; in the future, it will adapt media to serve the needs and tastes of each individual mind... Demoting reading from its privileged position in the school curriculum is only one of many consequences of Knowledge Machines...What follows from imagining a Knowledge Machine is a certainty that school will either change very radically or simply collapse. It is predictable that the education establishment cannot see farther than using new technologies to do what it has always done in the past, teach the same curriculum...The possibility of freely exploring worlds of knowledge calls into question the very idea of an administered curriculum."
- In a 1991 article for *The Whole Earth Review*, a quarterly magazine of access to tools and ideas, Roger Karraker discusses the Internet, quoting George Gilder. Karraker writes: "What would a real Network Nation be like? Conservative theorist/author George Gilder...foresees a renaissance in education...'The telecomputer could revitalize public education by bringing the best teachers in the country to classrooms everywhere,' Gilder says. 'More important, the telecomputer could encourage competition because it could make homeschooling both feasible and attractive. To learn social skills, neighborhood children could gather in micro-schools run by parents, churches, or other local institutions. The competition of homeschooling would either destroy the public school system or force it to become competitive with rival systems.'"
- Craig Lyndes was a teacher from Champlain Valley Union High School in Vermont who was a representative of his school in its membership in the National School Network Testbed (NSNT), funded by the National Science Foundation to encourage the effective promulgation of the Internet in

education. This statement was quoted in a research presentation made by Beverly Hunter, an NSNT official, at INET '95, the Internet Society's 1995 International Networking Conference. Lyndes (1995) reports: "Eventually we want to allow the students to access all of the school's resources from home. This is part of our long-range goal to blow the walls off the school, bring the world into the school, and put the school out in the world."

- In a 1993 article for *The Christian Science Monitor*, Romolo Gondolfo interviews Perelman, senior researcher at the Discovery Institute in Washington, D.C. Gondolofo quotes Perelman saying: "All bureaucracies, as we know, are rooted in the idea of controlling people's access to knowledge by concentrating it at the top and distributing it very parsimoniously to those at lower levels. But this is precisely what is becoming more and more difficult to do in this new Age of Knowledge which we are right now entering...A fundamental implication of this revolution is that the creation and transmission of knowledge will no longer move vertically, from the top down. It will move horizontally, among many people, at a tremendous speed. This will undermine the foundation of every bureaucracy, including schools...I propose to abolish all public grants for schools and colleges and instead give the money directly to families in the form of 'microvouchers' to be spent on anything that nurtures the spirit and teaches new skills" (p. 9).

- In his 1995 book, *Silicon Snake Oil*, writer Clifford Stoll shares his take on the Internet's future implications for education: "All of us want children to experience warmth, human interaction, the thrill of discovery, and solid grounding in essentials: reading, getting along with others, training in civic values. Only a teacher, live in the classroom, can bring about this inspiration. This can't happen over a speaker, a television or a computer screen. Yet everywhere I hear parents and principals clamoring for interactive computer instruction. What is wrong with this picture?...At the same time that school librarians, art instructors, and music teachers are being fired, we're spending thousands on computers. What's wrong with this picture?...'I believe that the motion picture is destined to revolutionize our educational system and that in a few years it will supplant largely, if not entirely, the use of textbooks.' —Thomas Edison, 1922. In the past, schools tried instructional filmstrips, movies, and television; some are still in use, but think of your own experience. Name three multimedia programs that actually inspired you. Now name three teachers that made a difference in your life" (p. 135).

PREDICTION ON DEMOCRATIC PROCESSES

> **SCENARIO:** *By 2014, network security concerns will be solved and more than half of American votes will be cast online, resulting in increased voter turnout.*

In the data in the Early 1990s Predictions Database, the general drift of the commentary splits in regard to this prediction when considering the concept of network security. Many people of the 1990s said security is not possible; some believed that a secure system would eventually be built; some said that a foolproof, secure system would ensure so much privacy that it would put the world's societies at the mercy of criminals and terrorists, shielding them from law enforcement; and some said that perfect encryption would make it impossible for governments to collect taxes. Here is a sample:

- In an interview for *InfoWorld* in 1994, Jayne Levin, editor of *The Internet Letter*, asks Daniel C. Lynch, "How can companies best protect their networks from intruders?" Lynch replies: "Ah, yes. Security. Networking. Let's see: secrecy and sharing. All together. Seems kinda contradictory. Network security will remain the hardest nut to crack (bad pun) for many years to come. Why? Because security itself is a perpetual problem. As long as we have humans around, anyway.

The Internet was created without much security included. We relied on the security of the individual operating systems residing on the hosts that were connected to the Internet. Then along came PCs and Macs. No concept of security was built into them...Many vendors and researchers are working on ways to extend the firewall concept to full-function Internet working. They will not be done tomorrow" (p. S72).

- In their 1994 book, *Firewalls and Internet Security*, William Cheswick and Steven Bellovin write: "It might seem that we are unduly pessimistic about the state of computer security. This is half-true. We are pessimistic, but not, we think, unduly so. Nothing in the recent history of either network security or software engineering gives us any reason to believe otherwise. Nor are we alone in feeling this way" (p. 8).
- In a 1994 article about digital democracy for *Wired* magazine, Evan I. Schwartz writes: "The very thought of living in an electronic democracy raises fundamental issues...Won't it be harder than ever for Congress and the President to stand up for what's right, rather than what's popular? Can voter privacy be maintained, or will marketers get hold of everyone's voting records? Will everyone have access to the latest technology? Will the people really be getting their say, or will the whole process by controlled by moguls like Malone? And perhaps most important, what would happen if votes somehow became binding, rather than just advisory?" (p. 75).
- In a 1995 essay for *Newsweek* magazine, Jonathan Alter quotes Neil Postman. Alter writes: "Although the technology already exists for a full-scale teledemocracy, no one has yet figured out a way to guarantee the integrity of the balloting. In fact, even computerized voting at polling places remains surprisingly suspect. 'The opportunities for rigging elections [are] child's play for vendors and knowledgeable election officials,' writes Peter G. Neumann in 'Computer-Related Risks.' (Neumann runs the Internet newsgroup *The Risks Forum*.) Short-term technical problems—like the disastrous pileup last November

in Canada when the Liberal Party tried a teleconvention with delegates voting from home by phone—can be fixed. But the larger problem of essentially turning over vote-counting to unaccountable computer experts will be unresolved for years. At least when Boss Tweed stole votes, everyone knew it. Computer vote fraud can be extraordinarily difficult to trace" (p. 34).

- In his 1994 book, *City of Bits*, MIT computer scientist William J. Mitchell writes: "As telecommunications networks have developed, there has been growing flirtation with the idea of replacing old-fashioned voting booths and ballot boxes with electronic polling. In a cyberspace election, you might find the policies of candidates posted online, you might use your personal computer to go to a virtual polling place to cast your vote, and the votes might be tallied automatically in real time...There are, of course, potential problems with electronic stuffing of ballot boxes, but these can be handled through password control of access to the virtual ballot box or (better) through use of encryption technology to verify a voter's identity...Electronic feedback can even be swift enough, potentially, to support real-time (or at least very fast) direct democracy on a large scale. Populist demagogues like Ross Perot have proffered visions of sitting in front of your two-way television, watching debates, and bypassing the politicians by immediately, electronically recording your response. The network presents the packaged alternatives. Vote with your remote!" (pp. 153–154).

- In his 1995 book, *Silicon Snake Oil*, writer Clifford Stoll shares his take on the Internet's future implications: "The myth holds that our networks are the ultimate in democracy—all voices can be heard. Bytes have no race, gender, age, or religion. What effects will we see when the government comes online? Computer access will let us send messages to government officials, and get quick responses from them. We'll know what's happening in the back rooms of our legislatures. We could read committees' reports the same day they're written and get fast responses to our queries. The myth grows: Elections will change, too. Politicians will be available through electronic forums, with less emphasis on expensive television ads. They'll upload position papers to the Net, and reply to e-mail from their constituents. Eventually, we'll see electronic voting—a way to further democratic participation, with polls giving near-instant feedback for representatives. The reality? Anyone can post messages to the Net. Practically everyone does. The resulting cacaphony drowns out serious discussion. Online debates of tough issues are often polarized by messages taking extreme positions. It's a great medium for trivia and hobbies, but not the place for reasoned, reflective judgment" (pp. 31–32).

- In an interview that aired on PBS–TV in 1995, Internet pioneer Stewart Brand said the following: "If total public cryptography and lots of financial transactions come to the Net, will you pay taxes in the future? You won't. This is one terrifying fantasy from the government standpoint. It may not be a fantasy, because if lots of transactions go onto the Net and they're completely encrypted in a way that they can't be tracked, a whole lot of financial activity basically goes black, goes underground. And then you can't tax transactions, you can't track transactions. All you've got left to tax basically is possessions at that point and so you may see...property taxes going up and sales taxes disappearing."

PREDICTION ON FAMILIES

> **SCENARIO:** *By 2014, as telework and home schooling expand, the boundaries between work and leisure will diminish significantly. This will sharply alter everyday family dynamics.*

In the data in the Early 1990s Predictions Database, the general drift of the commentary skewed in agreement with the prediction. While rosy statements were made at the time about the abundant choices to be made by families in a world of telecommuting and homeschooling, there were also notes of caution sounded about the impact of networked life in the home. Here are a few:

- In a 1995 article for *The Guardian* in London, Christopher Reed quotes *Wired* magazine editor/publisher Louis Rossetto. Reed writes: "What *Wired* is doing, in the words of cofounder and editor/publisher Louis Rossetto, is launching the 'digital revolution,' the 'creation and implementation of new electronic technology, what it means to our lives, and how it will change everything: business, politics, culture, education, art, and personal relationships.' The computers and the international networks, Rossetto believes, are media with such powerful messages that in a generation, the world will be a different place. Digitally doomed are mammoth corporations, political parties, the conventional school, the commute to the workplace, orthodox finances including national budgets, and popular entertainment...Even the family will change. 'What happens when families come back together because work is done at home?' asks Rossetto. 'What neuroses will that expose?'" (p. T14).
- In a 1995 online essay, Justin Hall makes the statement: "For the jobs of tomorrow, in the service sector, the home is the workplace. They already have a catchphrase for it—telecommuting...You can be near your kids, your pets, your garden, in the comfort of your own home, work on your computer and video teleconference to your meetings. Each office I have worked in has sucked up hours of my day. Dealing with other people's crises, lounging by the coffee machine, pointless meetings, getting from one place to the other. Home working, the time you waste is your own, around your family and friends. Set your own schedule, in your own environment, no commuting. If we abandon the concept of the inner-city office workplace, we can begin to unpave this country...people will be rooted in their local communities while maintaining global

presence. Home cooking and home improvement; the family structure will be bolstered by the presence of parents, in communities of energized folk" (Hall, 1995a).

- In his 1995 book, *Silicon Snake Oil*, writer Clifford Stoll shares his take on the Internet's future implications: "Networks hold out the promise of telecommuting. One day, many of us will be able to work at home, any hour of the day or night. We'll save gas, have closer family ties, and have a happier workplace. Oh? I doubt our offices will be replaced by minions working from home. The lack of meetings and personal interaction isolates workers and reduces loyalty. Nor is a house necessarily an efficient place to work, what with the constant interruptions and lack of office fixtures. Perhaps it'll work for jobs where one never has to meet anyone else, like data entry or telephone sales. What a way to turn a home into a prison" (p. 30).

- In a fall 1994 *New Perspectives Quarterly* article, "Magna Carta for the Knowledge Age," social critics Esther Dyson, George Gilder, Jay Keyworth, and Alvin Toffler write the following: "No one knows what the Third Wave communities of the future will look like, or where demassification will ultimately lead. It is clear, however, that cyberspace will play an important role in knitting together the diverse communities of tomorrow, facilitating the creation of electronic neighborhoods bound together not by geography but by shared interests. Socially, putting advanced computing in the hands of entire populations will alleviate pressure on highways, reduce air pollution, allow people to live further away from crowded or dangerous urban areas, and expand family time."

PREDICTION ON THE RISE OF EXTREME COMMUNITIES

> **SCENARIO:** *Groups of zealots in politics, in religion, and in groups advocating violence will solidify, and their numbers will increase by 2014 as tight personal networks flourish online.*

Future Scenarios

In the data in the Early 1990s Predictions Database, the commentary skews in agreement with the prediction. Of course, if the Internet can build up positive social networks, there is no reason to assume it could do anything but the identical thing for negative social networks. Here are a few selections:

- In a 1993 article for *Wired* magazine, futurist Peter Schwartz, a cofounder of the Global Business Network, discusses the high-tech future that will develop out of a knowledge-based world with futurist Alvin Toffler, the coauthor (with his wife, Heidi Toffler) of *Future Shock*, *The Third Wave*, and *War and Anti-War*. Schwartz quotes Toffler saying: "The world system is splitting into three parts—three different layers or tiers—or more accurately three different civilizations. Of course, you'll continue to have agrarian countries and you'll continue to have the mass-manufacturing, cheap-labor suppliers, at least for a transitional period. But we are…rapidly developing a chain of info-intensive countries whose economics depend not on the hoe or the assembly-line but on brainpower…The emerging third-wave civilization is going to collide head-on with the old first and second civilizations. One of the things we ought to learn from history is that when waves of change collide they create countercurrents. When the first and the second wave collided we had civil wars, upheavals, political revolutions, forced migrations. The master conflict of the 21st century will not be between cultures but between the three supercivilizations—between agrarianism and industrialism and post-industrialism."
- In a 1995 article in *Government Technology*, Blake Harris writes: "The dark side of cyberspace harbors hackers pirating software and exchanging hacking techniques, drug smugglers using e-mail, political extremists advocating racism, hate and violence, predators seeking to seduce children, pornographers with modems, and maybe even terrorist networks plotting atrocities—in fact, almost every form of evil that already exists in our society. It is a little terrifying, at times, to think

that virtually anyone, armed simply with a computer, modem, and telephone line, can, at least in theory, reach a worldwide audience with whatever communication he or she wishes. This fact, coupled with the anarchic freedom of the Internet, has brought to a head a number of fundamental issues that may have significant ramifications on how the Information Age unfolds: surveillance and public safety vs. privacy through encryption and anonymity, censorship vs. free pression, more control vs. a decentralized anarchy of information."

- In his 1994 book, *City of Bits*, MIT computer scientist William J. Mitchell writes: "Network pimps will offer ways to do something sordid (but safe) with lubriciously programmed telehookers. (This is an obvious extrapolation of the telephone's transformation of the whorehouse into the call-girl operation.) Telemolesters will lurk. Telethugs will reach out and punch someone" (pp. 19–20).

- In a 1995 article for *Computerworld*, Gary Anthes writes about the statements made by attendees at a recent conference on information warfare, quoting Stephen Kent of Bolt Beranek and Newman Inc. Anthes writes: "Problems and threats: 'It's clear that we have lots of vulnerable systems that the country depends on. Terrorist organizations are especially worrisome. They are eager for the kind of notoriety that would attend their knocking out the telephone system or air traffic control system.' —Stephen Kent, chief scientist for security technology at Bolt Beranek and Newman Inc., Cambridge, Mass" (p. 71).

- In a January 23, 1995, *U.S. News & World Report* article, "Policing Cyberspace," Vince Sussman explores First Amendment rights in cyberspace. The article includes an interview with FBI Special Agent William Tafoya. Sussman quotes Tafoya, writing: "'Crime involving high technology is going to be off the boards,' predicts FBI Special Agent William Tafoya, the man who created the bureau's home page on the Internet, the worldwide computer network. 'It won't be long before the bad guys outstrip our ability to keep up with them.'"

Sussman also interviews Carlton Fitzpatrick, branch chief of FLETC's Financial Fraud Institute. Sussman writes: "'Cyberspace is like a neighborhood without a police department,' says FLETC's [Carlton] Fitzpatrick. One of the most pressing dangers, says Fitzpatrick, is that people bound by hate and racism are no longer separated by time and distance. They can share their frustrations at nightly computerized meetings. 'What some people call hate crimes are going to increase, and the networks are going to feed them,' predicts Fitzpatrick, [branch chief of FLETC's Financial Fraud Institute]. 'I believe in the First Amendment. But sometimes it can be a noose society hangs itself with'" (pp. 55–60).

- In his 1994 book, *Out of Control: The New Biology of Machines, Social Systems and the Economic World*, Kevin Kelly, editor of *Wired* magazine, quotes Tim May in a discussion of the future impact of encryption and anonymous remailers: "I confess my misgivings about the potential market for anonymity to Tim: 'Seems like the perfect thing for ransom notes, extortion threats, bribes, blackmail, insider trading, and terrorism.' 'Well,' Tim answers, 'What about selling information that isn't viewed as legal, say about pot growing, do-it-yourself abortion, cryonics, or even peddling alternative medical information without a license? What about the anonymity wanted for whistleblowers, confessionals, and dating personals'" (Kelly, 1994b, p. 209).

- In a 1995 article in *New Scientist*, Kurt Kleiner reports on what Mike Godwin and David Banisar are saying about fears that the government may try to control or acquire the ability to tap into secure communications on the Internet. Kleiner writes: "Users of the Internet are afraid that there will be some sort of clampdown on them because of the wave of paranoia that has swept the country after the Oklahoma City bombing. Newspapers and TV shows have carried stories about the sort of information that is available over the Internet. For instance, they point out that *The Terrorist's Handbook* is easy to find, complete with detailed information on how to mix and detonate more than a dozen kinds of explosives, including the one used in the Oklahoma City bombing. It also became clear after the bombing that members of militia groups, such as the one the bomber belonged to, communicate via the Internet. Godwin points out that so far no one in a government position has called for censorship of the Internet. And David Banisar, of the Electronic Privacy Information Center, thinks such censorship is unlikely. 'What can they do? Say no political organizing over the Internet? That's clearly unconstitutional'" (p. 66).

- In May 1995 *Wired* magazine ran an article that was excerpted from a transcript of a speech Bruce Sterling delivered at the High Technology Crime Investigation Association conference in November 1994. Sterling says: "Countries that have offshore money laundries are gonna have offshore data laundries. Countries that now have lousy oppressive governments and smart, determined terrorist revolutionaries are gonna have lousy oppressive governments and smart determined terrorist revolutionaries with computers. Not too long after that, they're going to have tyrannical revolutionary governments run by zealots with computers; then we're likely to see just how close to Big Brother a government can really get. Dealing with these people is going to be a big problem for us" (p. 129).

PREDICTION ON POLITICS

> **SCENARIO:** *By 2014, most people will use the Internet in a way that filters out information that challenges their viewpoints on political and social issues. This will further polarize political discourse and make it difficult or impossible to develop meaningful consensus on public problems.*

In the data in the Early 1990s Predictions Database, some commentary is in agreement with this prediction. There were also some people who projected that the Internet could bring about a new pluralism. Here are some examples from both sides:

- For a 1995 article for *Wired* magazine, Jay Kinney, publisher and editor of *Gnosis: A Journal of the Western Inner Traditions*, writes: "One gets the sense that, given half a chance, the electorate would love to ditch the old left/right horseshoe match and take on some new paradigms altogether...Some techno-optimists, entranced with the rapid expansion of cyberspace, are convinced that the rough contours of the future can be spotted in the shadowy forms dancing across their computer screens. The pounding drums of cypherpunks, Usenet orators, civil-liberties activists, and venture capitalists, all undulating together in the flickering RGB glow, seem to whisper alluring promises of power, privacy, and pluralism in the politics to come...When all is said and done, is there a new politics emerging in the Net/cyberspace/digital culture? Short answer: Yes, if by 'new politics' one means an increased visibility for certain strains of ideology, like libertarianism, that have not generally made it through the mass media's bozo-filters. Libertarianism—with its zealous advocacy of laissez-faire capitalism, deregulation, and privatization—is a ready-made 'killer app' for high-tech start-ups, would-be millionaires, and the rest of the 'don't tread on me' cybercrew. Mix this in with the current impatience toward half-failed liberal solutions and mammoth government and we may see some unusually radical proposals enacted in Washington" (pp. 90, 95).
- A paper titled "Computer-Mediated Communication and the American Collectivity: The Dimensions of Community Within Cyberspace," by Jan Fernback and Brad Thompson, was presented at the annual convention of the International Communication Association, Albuquerque, New Mexico, in May 1995. It was reprinted in full form on Howard Rheingold's Web site. This is an excerpt: "CMC [computer-mediated communication] does not, at this point, hold the promise of enhancing

democracy because it promotes communities of interest that are just as narrowly defined as current public factions defined by identity (whether it be racial, sexual, or religious). Public discourse ends when identities become the last, unyielding basis for argumentation that strives ideally to achieve consensus based on a common good."

- In a 1994 article for *Wilson Quarterly*, Edward Tenner writes: "Yes, networks can help people strengthen neighborhoods and communities. But they also encourage people to find ways out. Unhappy with your schools? Join the parents who have turned to homeschooling. Teaching materials and mutual support are already available online, and home educators have been using electronic mail effectively to organize and lobby for their rights. Their children may learn all they need to, but the economist Albert O. Hirschman has pointed out that when the most quality-conscious users are free to leave a troubled system, whether railroads or schools, the system suffers further by losing its most vocal critics. Any future information network will help unhappy people secede, at least mentally, from institutions they do not like, much as the interstate highway system allowed the affluent to flee the cities for the suburbs and exurbs. Prescribing mobility, whether automotive or electronic, as an antidote to society's fragmentation is like recommending champagne as a hangover remedy" (p. 37).

- In a 1995 article in *Le Monde Diplomatique*, Paul Virilio, the emblematic French theorist of technology and author of *Pure War, Speed and Politics*, and *War and Cinema: The Logistics of Perception*, writes: "The dictatorship of speed at the limit will increasingly clash with representative democracy. When some essayists address us in terms of 'cyber-democracy,' of virtual democracy; when others state that 'opinion democracy' is going to replace 'political parties democracy,' one cannot fail to see anything but this loss of orientation in matters political, of which the March 1994 'media-coup' by Mr. Silvio Berlusconi was an Italian-style prefiguration. The advent of the age of viewer-counts and opinion polls reigning supreme will necessarily be advanced by this type of technology."

- In a 1995 article for *Governing*, Christopher Conte quotes Andrew Blau of the Benton Foundation. Conte writes: "The ease and sheer speed of new communications technologies give the information superhighway its appeal. But those same qualities provoke fears among people who see it as a threat to representative democracy. 'Real democracy is slow and deliberative,' notes Andrew Blau, director of the communications policy project for the Benton Foundation, a Washington, D.C., group that promotes the use of the information superhighway. 'It's so easy to imagine a scenario in which technology is used to get instant judgments from people. If it is used that way, we haven't seen anything yet when it comes to high-tech lynchings'" (p. 34).

- In his 1995 book, *The Electronic Republic: Reshaping Democracy in the Information Age*, Lawrence Grossman, former president of NBC News and PBS, writes: "Boiler room' organizations, hired by special interests, will seek to manufacture and mobilize 'grassroots' opinion and stimulate the outpouring of selected messages and votes to make sure that particular viewpoints are heard. They do that now. In the next century, it will become a mainstream business. Computerized political advertising, promotion, and marketing campaigns, targeted with as much intensity as legislators, regulators, and public officials are lobbied today, because public opinion—the fourth branch of government—will play an even more pivotal role in major government decisions" (p. 150).

- In the July 1994 issue of *The Network Observer* online newsletter, Barbara Welling Hall writes about networking, democracy, and computers: "Inequitable access to these technologies at present and in the foreseeable future profoundly diminishes the diversity of opinions that are vetted electronically. Electronic communities may provide genuine benefits to isolated individuals, but if these communities are to be presented as providing global rather than partial access to political discourse, this promise may be squandered. Finally, although freedom of information may hamper some dangerous actions, more information alone is not a substitute for the development of critical or compassionate faculties. Data may reveal the existence of injustice, but data alone rarely generate the political will either to make difficult trade-offs or to discover creative solutions to perennial problems."

Prediction on Health System Change

> **Scenario:** *In 10 years, the increasing use of online medical resources will yield substantial improvement in many of the pervasive problems now facing health care—including rising health care costs, poor customer service, the high prevalence of medical mistakes, malpractice concerns, and lack of access to medical care for many Americans.*

In the data in the Early 1990s Predictions Database, the general drift of the commentary skews in agreement with the prediction. The Internet was seen as a way to streamline the connection between consumers and health care providers, and the availability of health education data was seen as a key to improving millions of lives in Third World nations. Here are some examples:

- In a 1994 article for *Wired* magazine, Joe Flower explains the types of changes that could come in health care through the use of networked computing. Flower writes: "The coming American health care system has everything to do with smart cards and dumb terminals, big bandwidth and microprobes, genetic markers and info-markets. And it doesn't look like anything you've read in the paper...Over the next decade, we will see health care become less doctor-centered, and more community and family-centered. Medicine itself will become less of an art and more fact-based. Yet at the same time it will come to feel more humane...Eventually all health records, from insurance information to X-rays and MRI scans will go digital—and eventually you will carry all that information with you on a card...The very discoveries and inventions that will continue to transform medical practice will push it to be less about hardware, less about vast and powerful machines watched over by highly trained acolytes, and more about shared information... They carry the possibility of providing major assistance in revolutionizing health care, making it both cheaper and better, spreading it wider, involving people in making decisions about their own lives, helping America (and eventually the world) build truly healthier communities" (pp. 108, 110, 112–113, 150).

- In a 1993 report of the Information Infrastructure Task Force, "The National Information Infrastructure: Agenda for Action," members of the commission report as follows: "Experts estimate that telecommunications applications could reduce health care costs by $36 to $100 billion each year while improving quality and increasing access. [The following] are some of the existing and potential applications. —Telemedicine: By using telemedicine, doctors and other care givers can consult with specialists thousands of miles away, continually upgrade their education and skills, and share medical records and X-rays... —Unified Electronic Claims: More than 4 billion health care claims are submitted annually from health care providers to reimbursement organizations such as insurance companies, Medicare, Medicaid, and HMOs. The administrative costs of the U.S. health care systems could be dramatically reduced by moving towards standardized electronic submission and processing of claims. —Personal Health Information Systems: The United States can use computers and networks to promote self-care and prevention by making health care information available 24 hours a day in a form that aids decision making. Michael McDonald, chairman of the Communications and computer Applications in Public Health (CCAPH) estimates that even if personal health information systems were used only 25 to 35% of the time, $40 to $60 billion could be saved. —Computer-Based Patient Records: Computer-Based Patient Records are critical to improving the quality and reducing the cost of health care."

- A research group representing The Global Health Network, an international group with the hope of using the Internet to establish a better world of medicine and prevention, made the following statement in a research presentation at INET '95, the Internet Society's 1995 International Networking Conference. The group reports: "We should be able to monitor and forecast diseases as well as we monitor the weather if we take on new technologies. Having an Internet backbone to national and global-disease monitoring can yield accurate and timely information concerning disease conditions...At the time of a disaster one of the needs—if not the most critical—is that of communication" (LaPorte, Villasenor, & Gamboa, 1995).

PREDICTION ON THE PERSONAL ENTERTAINMENT AND MEDIA ENVIRONMENT

> SCENARIO: *By 2014, all media, including audio, video, print, and voice, will stream in and out of the home or office via the Internet. Computers that coordinate and control video games, audio, and video will become the centerpiece of the living room and will link to networked devices around the household, replacing the television's central place in the home.*

In the data in the Early 1990s Predictions Database, the general drift of the commentary skews in agreement with this prediction. At the time, whether they called it a television, a computer or a teleputer, most experts were saying that our homes would be tied to a network of information flowing inward and outward. Here are some examples:

- In a 1994 article for *Wilson Quarterly,* Douglas Gomery writes: "The basic device serving consumers at home will almost certainly be some sort of hybrid telecomputer that marries a computer processor and a television screen. It will display wide-screen images, easily accommodating all of Hollywood's

CinemaScope-like images without lopping off the sides... Telecomputers of the sort described here will cost thousands of dollars each. When they finally become widely available, for example, digital high-definition television (HDTV) sets are likely to cost in the neighborhood of $5,000. To wire the nation with fiber-optic cable, add at least $1,000 per household, or a cool $100 billion for the whole country. That is not to mention the cost of wiring businesses, government, and nonprofit institutions. Sums of this size serve as reminders that, much as we like to think of the infohighway as the centerpiece of a 'postindustrial' era, building it will be a very old-fashioned capital-intensive undertaking. It will take a long time, and it will be very expensive" (p. 8).

- In a 1995 article for *Time*, reporter Barrett Seaman writes about future technologies. He quotes Mark Weiser. He writes: "At Xerox's Palo Alto Research Center in California, where the PC, on-screen icons and the laser printer originated, Mark Weiser, manager of the computer science laboratory, envisions a world in which flat-panel screens bearing a multitude of images will be household regulars. They will range from tiny ones, costing perhaps $5 each and plastered everywhere, to wall-size ones for viewing video. The smaller ones, says Weiser, are 'where you'll plan your grocery list or do your homework. They'll be the equivalent of Post-it notes on the refrigerator or the crumpled-up notepaper in your pocket.' In Weiser's world, people will wake up to a tiny bedside screen that gives the time and the weather forecast and even displays news headlines or sports scores. Pocket-size screens would also serve as remote controls for larger screens in the bedroom or living room, where family members will use them variously to watch TV, read the newspaper (which will be customized for each member's personal interests), or draw up the family grocery list" (p. 33).
- In his 1995 book, *The Road Ahead*, Microsoft CEO Bill Gates writes: "Your television set will not look like a computer and won't have a keyboard, but the additional electronics inside

or attached will make it architecturally a computer like a PC. Television sets will connect to the highway via a set-top box similar to ones supplied today by most cable TV companies. But these new set-top boxes will include a very powerful general-purpose computer. The box may be located inside a television, behind a television, on top of a television, on a basement wall, or even outside the house. Both the PC and the set-top box will connect to the information highway and conduct a 'dialogue' with the switches and servers of the network, retrieving information and programming and relaying the subscriber's choices" (p. 70).

- In a 1994 article he wrote for *National Re*view, George Gilder, a Fellow of the Discovery Institute in Seattle and author of *Life After Television*, expounds on his views of future communications. He writes: "Within the next 10 years, this explosive technological advance in both networks and processors virtually guarantees that the personal-computer model of distributed intelligence and control will unseat the emperors of the mass media and blow away the television model of centralization. The teleputer—a revolutionary PC of the next decade—will give every household hacker the productive potential of a factory czar of the industrial era and the communications power of a broadcast tycoon of the television age. Broadcasting hierarchies will give way to computer heterarchies—peer networks in which the terminals are essentially equal in power and there is no center at all" (Gilder, 1994a).

PREDICTION ON CREATIVITY

> **SCENARIO:** *Pervasive high-speed information networks will usher in an Age of Creativity in which people use the Internet to collaborate with others and take advantage of digital libraries to make more music, art, and literature. A large body of independently produced creative works will be freely circulated online and will command widespread attention from the public.*

In the data in the Early 1990s Predictions Database, the commentary followed in agreement with this prediction. At the time, the digital libraries were mostly in the early planning stages, but the Internet was born from the desire for human collaboration, so it was natural to assume that creative people would be able to mimic the successes found in collaborations on the Internet by researchers in the network community. Here are some examples:

- In an excerpt from his 1994 book, *Life After Television*, George Gilder addresses the future: "The new law of networks exalts the smallest coherent system: the individual human mind and spirit. A healthy culture reflects not the psychology of crowds but the creativity and inspiration of millions of individuals reaching for high goals. In place of the broadcast pyramid, a peer network will emerge in which all the terminals will be smart" (Gilder, 1994b, pp. 49–50).
- In a 1994 article for *Wired* magazine, Daniel Pinchbeck, a New York-based writer and the editor of *Open City*, a literary and art journal, writes: "Eventually, computers and the Internet may force artists out of the increasingly esoteric discourse of the art world. A broader audience may demand that they reintegrate their work with larger issues related to science, technology, and humanism. 'I would like to see a return to that classical breadth of inquiry that artists were able to make in the Renaissance,' says Michael Joaquin Grey. Computers may also force radical artists to return to a notion of craft. In the contemporary art world, painstaking studio process often seems to matter less than an up-to-the-minute ironic pose. Artists of the past had to grapple with techniques ranging from draftsmanship to fresco painting if they wanted to achieve greatness. Their creative inheritors may have to master digital tools if they hope to reach beyond the restrictive walls of galleries and museums" (p. 208).
- In his 1995 book, *The Road Ahead*, Microsoft CEO Bill Gates writes: "Over time we will start to *create* new forms and formats that will go significantly beyond what we know now. The exponential expansion of computing power will keep changing the rules and opening new possibilities that

will seem as remote and farfetched then as some of the things I've speculated on here might seem today. Talent and creativity have always shaped advances in unpredictable ways...The information highway will open undreamed-of artistic and scientific opportunities to a new generation of geniuses" (p. 134).

- In a 1991 article for *The New York Times*, John Markoff interviews Internet pioneer Robert Kahn as he explains the nation's planned "national data highway." Markoff writes: "This network of fiber-optic cables, which would completed replace existing copper lines, is viewed by many scientists and executives as both a vital research tool and an essential part of the country's 'information infrastructure' for the next century...' [The Internet] will unleash a tremendous amount of creativity and innovation which will lead to capabilities we can't even imagine today,' said Robert Kahn, a scientist at the Corporation for National Research Initiatives, a Reston, Va., research organization that is coordinating consortiums of corporations, research laboratories and universities developing extremely fast computer networks" (p. 39).
- In a 1995 article for *Wired* magazine, media critic Jon Katz writes: "The explosion of energy coming from digital designers, musicians, filmmakers, photographers, and even advertisers is altering our basic notions of creativity. A new dream of the future is being born. Of course, in a half-century or so, these same digital revolutionaries will form the nostalgic material of somebody else's 'history.' Imagine the writer of that book, or CD-ROM, or digital bedside laptop tablet, longing for the time when clunky computers sprouted wires, modems hissed, and chips held finite memory. Think how much wonder our time might hold" (p. 86).

PREDICTION ABOUT HOW PEOPLE GO ONLINE

SCENARIO: *By 2014, 90% of all Americans will go online from home via high-speed networks that are dramatically faster than today's high-speed networks.*

In the Early 1990s Predictions Database, the commentary was firmly behind the development of high-speed networks, and the vision was that these would come sooner rather than later. Most of this came after the introduction of the World Wide Web, with its ability to display images; the prospect of being able to exchange large video files and stream live video programming drove the development of this network technology. Here are some statements on delivering large amounts of data at high speeds:

- In a 1993 paper in reaction to the National Telecommunications and Information Administration's proposed goals, New York Law School Professor Michael Botein writes: "This area is fraught with perils…the U.S. government and industry should proceed with caution in entering the new digital age. High-capacity fiber systems probably will become a mainstay of all developed countries' telecom infrastructure at some point in the next millennium. For the moment, however, it might be wise to slow the whole process down a bit. After all, the U.S. should know by now that being first into a new technology is not always a benefit; the…experience with NTSC television should be a nightly reminder to U.S. telecom policy planners of the value of letting others make the mistakes."
- In a 1995 article for *Computerworld*, Gordon Bell looks ahead. Bell proposed a plan for a U.S. research and education network in a 1987 report to the Office of Science and Technology in response to a congressional request by Al Gore. He was a technology leader at Digital Equipment Corporation (where he led the development of the VAX computer) and with Microsoft: "Phone communications will evolve toward a single, pervasive digital dial tone for high-speed networks. These will offer bandwidth scalable to several hundred megabits per second for handling video over phone lines and virtual reality" (p. 89).
- In his 1994 book, *City of Bits*, MIT computer scientist William J. Mitchell writes: "The bondage of bandwidth is displacing

the tyranny of distance, and a new economy of land use and transportation is emerging—an economy in which high-bandwidth connectivity is an increasingly crucial variable... The most crucial task before us is not one of putting in place the digital plumbing of broadband communications links and associated electronic appliances (which we will certainly get anyway), nor even of producing electronically deliverable 'content,' but rather one of imagining and creating digitally mediated environments for the kinds of lives that we will want to lead and the sorts of communities that we will want to have" (pp. 5, 17).

- In a 1995 article for *Wired* magazine, Nicholas Negroponte, founder of MIT's Media Lab, writes: "In 2020, people will look back and be mighty annoyed by our profligate insistence on wiring a fiber-coax hybrid to the home rather than swallowing the cost of an all-fiber solution. They'll ask, 'Why didn't our parents and grandparents plan more effectively for the future?' As far as the American home is concerned, the phone companies have the right architecture (switched services), and the cable companies have the right bandwidth (broadband services). We need the union of these: switched broadband services. But how do we get from here to there? No one will deny that the long-term solution is to install fiber all the way, but the benefits seem diffuse and the costs acute. In the eyes of the telcos and cable companies, the question is financial—and since the near-term balance sheets don't add up, fiber is not being laid all the way. One way around this problem is to circumvent the private market and let a telecommunications monopoly build the infrastructure, which is exactly what Telecom Italia is doing...Italy will have a far better multimedia telecommunications system than the United States by 2000" (p. 220).

REFERENCES

Alter, J. (1995, February 27). The couch potato vote: Soon you'll be able to vote from home—but should you? *Newsweek*.

Anthes, G. (1995, October 2). Security pundits weigh war threat. *Computerworld, 29*(40), 71.

Barlow, J. P. (1994, March). The economy of ideas. *Wired, 2*(3), 84–90.

Bell, G. (1995, March 20). The view from here. *Computerworld, 29*(12), 88–89, 92.

Botein, M. (1993). *A few simple questions, with few good answers*. Elon University/Pew Internet Project. Retrieved December 20, 2007, from the Imagining the Internet: A History and Forecast Web site: http://www.elon.edu/predictions/prediction2.aspx?id=LCB-0012

Brand, S. (1995, June 15). *PBS interview with Stewart Brand*. Retrieved December 5, 2007, from http://www.pbs.org/wgbh/pages/frontline/cyberspace/ brand.html

Cheswick, W. R., & Bellovin, S. M. *Firewalls and Internet security: Repelling the wily hacker*. Indianapolis, IA: Addison-Wesley Professional.

Conte, C. (1995, June). Teledemocracy: For better or worse. *Governing, 8*(9), 33–37.

Creedy, S. (1995, May 2). Nurturing the Net. *Pittsburgh Post-Gazette*, p. C6.

Doheny-Farina, S. (1994, October). The last link: Default = offline or why ubicomp scares me. *Computer-Mediated Communication, 1*(6), 18.

Dyson, E. (1995, July). Intellectual value. *Wired, 3*(7), 136.

Dyson, E., Gilder, G., Keyworth, J., & Toffler, A. (1994, fall). A Magna Carta for the knowledge age. *New Perspectives Quarterly, 11*(4).

Elmer-Dewitt, P. (1994, July 25). Battle for the soul of the Internet. *Time, 144*(4), 50–56.

Fernback, J., & Thompson, B. (1995, May). *Computer-mediated communication and the American collectivity: The dimensions of community within cyberspace.* Paper presented at the Annual Convention of the International Communication Association, Albuquerque, New Mexico. Retrieved December 5, 2007, from http://www.rheingold.com/texts/techpolitix/VCcivil.html

Flower, J. (1994, January). The other revolution in health care. *Wired, 2*(1), 108–113, 150.

Gates, B. (1995). *The road ahead.* New York: Penguin Books.

Gilder, G. (1994a, August 14). Breaking the box. *National Review, 46*(15), 37–43.

Gilder, G. (1994b). *Life after television.* New York: W.W. Norton & Company.

Godwin, M. (1993, September/October). The law of the net: Problems and prospects. *Internet World*, 52–54.

Gomery, D. (1994, summer). In search of the cybermarket. *The Wilson Quarterly, 18*(3), 8.

Gondolfo, R. (1993, September 22). Will technology alter traditional teaching? *Christian Science Monitor*, p. 9.

Grossman, L. K. (1995). *The electronic republic.* New York: Viking.

Guglielmo, C. (1993, November/December). Man with a plan: Ed Lyell has actually been elected to the Colorado State Board of Education. *Wired, 1*(6).

Hall, B. W. (1994, July). Electronic networking and democracy. *The Network Observer, 1*(7). Retrieved December 5, 2007, from http://polaris.gseis.ucla.edu/pagre/tno/july-1994.html#electronic

Hall, J. A. (1995a, July 21). *Computopia: Sharing stories humanizes computer connections.* Retrieved December 5, 2007, from http://www.links.net/dox/tech/computopia.html

Hall, J. A. (1995b, July 23). *I am not my habits: On our guard against targeted advertising*. Retrieved December 5, 2007, from http://www.links.net/dox/tech/targad.html

Harris, B. (1995, October 1). Cyberspace 2020. *Government Technology*. Retrieved December 5, 2007, from https://www.govtech.com/gt/96050

Information Infrastructure Task Force. (1993). *The national information infrastructure: An agenda for action*. Washington, DC: Department of Commerce.

Karraker, R. (1991, spring). Highways of the mind or toll roads between information castles? *Whole Earth Review, 70*, 4–11.

Katz, J. (1995, October). Lost world of the future: Looking back at the 1939 New York World's Fair, David Gelernter's "novel with an index" exposes the irrevocable link between technology and nostalgia. *Wired*, 3(10), 82–86.

Kelly, K. (1994a, June 20). In 2004 we'll all live on the Internet with Silicon Valley visionaries. Kevin Kelly already does. *The Guardian (London)*, p. S2.

Kelly, K. (1994b). *Out of control: The new biology of machines, social systems, and the economic world*. Reading, MA: Addison-Wesley.

Kinney, J. (1995, September). "Anarcho-Emergentist-Republicans": Is there a new politics emerging in the net/cyberspace/digital culture? *Wired*, 3(9), 90–95.

Kleiner, K. (1995, May 6). Electronic vigilantes win friends after Oklahoma bombing. *New Scientist*, 66.

LaPorte, R., Villasenor, A., & Gamboa, C. (1995, June). *The Global Health Network*. Presented at the ISOC INET '95 Conference, Honolulu, HI.

Lawrence, N. (1994, March). Shocking new assaults on your privacy: Big Brother is watching, and he's probably got your number. *Midwest Today*. Retrieved December 5, 2007, from http://www.midtod.com/highlights/privacy.phtml

Levin, J. (1994, May 2). To dream the Internetworking dream. *Infoworld, 16*(18), S72.

Lewis, P. (1995, September 4). Computers beware! New type of virus is loose on the Net. *The New York Times*, p. 1.

Lyndes, C. (1995, June). *Internetworking and educational reform: The national school network testbed.* Presented at the ISOC INET '95 Conference, Honolulu, HI.

Magnish, S. (1994, June 5). Lawless cyber frontier. *Toronto Sun*, p. 51.

Markoff, J. (1991, January 1). Transforming the decade: 10 critical technologies; Fiber optics new networks for the nation. *The New York Times*, p. 39.

Mitchell, W. J. (1994). *City of bits.* Cambridge, MA: The MIT Press.

Negroponte, N. (1995, October). 2020: The fiber-coax legacy: What we see in the current fiber-coax strategies is fiscal timidity, justified by the usage patterns of an old-line broadcast and publishing model, not the Net. *Wired, 3*(10), 220.

NII Copyright Protection Act: Hearing before the Senate Judiciary Committee and the House Committee on Judiciary Subcommittee on Courts and Intellectual Property, S. 1284, H.R. 2441, 104th Cong. (1995) (testimony of Sen. Patrick Leahy). Federal News Service. Retrieved from the LexisNexis database.

Papert, S. (1993, May/June). Obsolete skill set: The 3 Rs. *Wired*, 1(2).

Pinchbeck, D. (1994, December). State of the art. *Wired*, 2(12), 156–159, 206–208.

Reed, C. (1995, March 20). Inter next world; The medium is the message in the case of. *The Guardian (London)*, p. T14.

Rheingold, H. (1993). Cold knowledge and social warmth. *Newsweek, 10*(49).

Rose, K. (1995, June). Learning with the World Wide Web: Connectivity alone will not save education. Presented at the ISOC INET '95 Conference, Honolulu, HI.

Rothfeder, J. (1992). *Privacy for sale.* New York: Simon & Schuster.

Sale, K. (1995). Setting limits on technology. *The Nation, 260*(22), 785–788.

Scheer, C. (1995). The pursuit of techno-happiness. *The Nation, 260*(18), 632–634.

Schwartz, E. I. (1994, January). Direct democracy: Are you ready for the Democracy Channel? *Wired, 2*(1), 74–75.

Schwartz, P. (1993, November). Shock wave (anti) warrior. *Wired, 1*(5).

Schwartz, P. (1995, April). Warrior in the age of intelligent machines. *Wired, 3*(4), 138.

Sclove, R. (1995). *Democracy and technology.* New York: The Guilford Press.

Seaman, B. (1995, spring). The future is already here. *Time, 145*(12), 30–33.

Sterling, B. (1995, May). Good cop, bad hacker: Bruce Sterling has a "frank chat" with some cops. *Wired, 3*(5), 122–129.

Stoll, C. (1995). *Silicon snake oil.* New York: Doubleday.

Sussman, V. (1995, January 23). *U.S. News & World Report,* 55–60.

Tenner, E. (1994). Learning from the Net. *The Wilson Quarterly, 18*(3), 37.

Virilio, P. (1995, August). Speed and information: Cyberspace alarm! (P. Riemens, Trans.). *CTheory.* (Reprinted from *Le Monde Diplomatique.*) Retrieved December 6, 2007, from http://www.ctheory.net/articles.aspx?id=72

Whalen, J. (1995, March). You're not paranoid: They really are watching you. *Wired, 3*(3), 76–85.

Appendix II

Biographies

The 1,286 participants in the survey were allowed to retain complete anonymity or they could enter their names while retaining the right to keep their answers anonymous; many longtime Internet luminaries chose to remain completely anonymous, and their names are not in the following compilation, nor are they used in any aspect of the final report. The survey respondents also had the opportunity to elect on each and every question they were asked whether or not they chose to have their name tied to their answer.

The following list of more than 100 biographies is a sample of respondents. It is not all-inclusive. It is offered to give a brief illustration of the expertise and background typical of most of the participants who were willing to give their names. Please note that many companies/groups/agencies were represented by multiple participants; if you see a quote credited in this site's many pages of online documentation of the study without a name listed but rather just a workplace (such as Microsoft, Media General or the FCC), remember that more than one person from that place took part in the survey, so that expressed view won't necessarily represent the view of any of the people whose names you see listed elsewhere in this online report about the survey.

Among the Future of the Internet I survey respondents who were willing to forgo total anonymity were the following people:

Carol Adams-Means, University of Texas–San Antonio: Adams-Means researches the social migration of special populations to an information society, telecommunications policy, the use of new media by minorities, and the use of new-media strategies in education and business.

Lois C. Ambash, Metaforix Inc.: Ambash is from New York, and in addition to being president of her own Web PR company, she serves on the board of the Internet Healthcare Coalition and on URAC's Health Web Site Accreditation Committee. She is a columnist for LLRX.com (a law and technology Web site) and a contributor to 2young2retire.com (a site for people seeking alternatives to retirement), and also contributes to other print and online venues.

Gary Arlen, Arlen Communications Inc., Alwyn Group LLC: Arlen is president of his Bethesda, Maryland-based research and consulting firm that specializes in interactive program content. He is known for his insights into the development of applications, especially interactive content for Internet, two-way TV, and other emerging systems.

Reid Ashe, Media General, Inc.: Ashe is the president and chief operating officer of the most-converged news media company in the United States; Media General's Tampa news operation, for instance, has its TV station, newspaper, and online unit all sharing the same space and resources. Ashe had previously been an executive with Knight Ridder.

Rob Atkinson, Progressive Policy Institute: Atkinson is vice president of this think tank and the director of its "Technology and New Economy Project." He was previously project director at the Congressional Office of Technology Assessment, and in 1995 he directed "The Technological Reshaping of Metropolitan America." He is a board member or advisory council member of the Alliance for Public Technology, Information Policy Institute, Internet Education Foundation, NanoBusiness Alliance, and NetChoice Coalition. He also serves on the advisory panel to Americans for Computer Privacy.

Gary Bachula, Internet2: Bachula is vice president for external relations for Internet2. Before that he was acting under secretary of Commerce for Technology at the U.S. Department of Commerce, where he led the formation of government-industry partnerships.

Biographies

Paul M. A. Baker, Georgia Centers for Advanced Telecommunications Technology: Baker is a senior research scientist and the Wireless RERC project director of "Policy Initiatives" to support universal access. He is also an affiliate assistant professor at the George Mason University School of Public Policy in Fairfax, Virginia. His areas of research interest include public-sector information-policy development and state and local government use of information and communication technologies (ICTs).

Troy Barker, ICF Consulting: Barker works with a management, technology, and policy consulting firm that develops solutions in regard to energy, environment, homeland security, community development, and transportation.

Jordi Barrat i Esteve, Electronic Voting Observatory, Universitat Rovira i Virgili: Barrat i Esteve's research is concentrated on e-voting. His university is located in Tarragona, Spain.

Christine Boese, CNN Headline News: Boese is also a cyberculture researcher and a columnist for CNN.com. She has also worked as a consultant and a college professor. Her areas of research interest include cyberculture studies, Weblogs, social network computing, interaction architecture, and hypermedia and multimedia communication theory.

Bill Booher, Booher & Associates: Booher served as deputy assistant secretary for Technology Policy at the U.S. Department of Commerce and currently is chief of staff at the National Telecommunications and Information Administration, George H. W. Bush's policy advisory group on domestic and international telecommunications issues.

Mike Botein, Media Center, New York Law School: Botein is also one of the voices in the Early 1990s Predictions Database. He was founding director of the Communications Media Center at New York University Law School. His expertise in international telecommunications law and the regulation of cable television and new technologies have made him a valuable consultant to the FCC and the Administrative Conference of the United States. He wrote *International Telecommunications in the United States* and *Cases and Materials on Regulation of the Electronic Mass Media.*

James Brancheau, vice president, Gartner Research: Brancheau's experience includes work as an entrepreneur and founding principal of Solista Global—a global emerging-technology consulting firm, managing director of Solista-Europe, lead planner for major ITV launch, editor of O'Reilly learning series on Web applications development, CIO in the higher-education sector, software engineer, and Web designer. He has been active in research centered on network technology since the early 1990s. He is an expert on the speed of adoption of new technologies and the new digital home.

Bradford C. Brown, National Center for Technology and Law: Brown is a columnist for *InformationWeek* and serves as chairman of the National Center for Technology and Law at the George Mason University School of Law. Topics of his columns have ranged from RFID to e-voting.

Jonathan Band, partner, Morrison & Foerster LLP: Band's work for this Washington, D.C.-based law firm is concentrated in the areas of intellectual property, computers and software, privacy and Internet, and new media.

Laura Breeden, director, America Connects Consortium, Education Development Center—established by the U.S. Department of Education in 2000 to strengthen community technology centers: Previously, she was an independent consultant focusing on Internet strategies and organizational development. Her clients included SRI International, the Morino Institute, the James Irvine Foundation, and other leading institutions that study, develop, and promote the use of network technologies. From 1994 to 1996, Breeden was director of a highly competitive, multi-million-dollar federal grant program (now known as TOP) designed to demonstrate the benefits of the "information superhighway" in the public sector. Under her leadership, more than 200 organizations received a total of $60 million for innovative community projects.

Michael Buerger, Bowling Green University, Police Futurists International, Futures Working Group: Buerger has been a Visiting Fellow at the National Institute of Justice, served as director of the Minneapolis Office of the Crime Control Institute, and was research director for the Jersey City (New Jersey) Police Department. He is a charter member of the Futures Working Group (FWG), a collaborative agreement between Police Futurists International and the Federal Bureau of Investigation. He is coauthor of an FWG white paper on augmented reality (AR) systems for police.

Kate Carruthers, Carruthers Consulting: Carruthers' consulting business is based in Australia, where she previously worked with the New South Wales Government as program director. She is a former chair of the Institute of Electrical & Electronics Engineers (IEEE) Computer Society in NSW. She is on the steering committee for Females in Information Technology & Communications.

Eliot Chabot, senior systems analyst, House Information Resources: Chabot works for the information technology office for the U.S. House of Representatives.

Gary Chapman, University of Texas at Austin: Chapman is director of the 21st Century Project at the graduate school for public policy at the University of Texas. He was executive director of Computer Professionals for Social Responsibility from 1984 to 1991, he was director of CPSR's 21st Century Project from 1991 to 1993, and he edited the CPSR newsletter from 1985 to 1993. His research has been funded for years by the National Science Foundation. He also wrote an internationally syndicated biweekly newspaper column on technology and society named "Digital Nation" for 6 years; it was published and syndicated by the *Los Angeles Times*. He has served on the selection committee for the Turing Award—the computer-science field's equivalent of the Nobel Prize, and chaired the five-member committee in 2004.

Stanley Chodorow, professor emeritus, University of California–San Diego: Chodorow is a historian who became the founding chief executive of the California Virtual University, a consortium of accredited institutions of higher education that offer distance-learning programs. He was provost of the University of Pennsylvania from 1994 to 1997. He is a board member with the Council on Library and Information Resources in Washington, D.C., and the Center for Research Libraries. He has also been an executive with Questia Media Inc., an online -information-resources company.

Ted Christensen, coordinator, Arizona Regents University: Christensen coordinates the development of e-learning at Arizona's three public universities: Arizona State University, Northern Arizona University, and the University of Arizona.

Thomas Claburn, *InformationWeek*: Claburn is a writer and editor at *InformationWeek* and formerly worked at *New Architect*, *Wired*, and KQED-TV.

Ben Compaine, communications policy expert: Compaine is editor of the book *The Digital Divide: Facing a Crisis or Creating a Myth?* and is coauthor of *Who Owns the Media?* He is a telecommunications expert and worked as a consultant for the MIT Program on Internet and Telecoms Convergence.

Noshir Contractor, University of Illinois at Urbana-Champaign: Contractor is the principal investigator on a 3-year, $1.5 million grant from National Science Foundation's Knowledge and Distributed Intelligence Initiative to study the coevolution of knowledge networks and 21st-century organizational forms. Previously, he was a coprincipal investigator on the National Science Foundation's Project CITY (Civil Info-structure Technology), which examined infrastructure development and maintenance. His research has also been supported, in part, by grants from the Sloan Foundation, the Annenberg Foundation, the U.S. Department of Education, and corporate sponsors including Apple, 3M, Steelcase, and Panasonic.

Biographies

Susan Crawford, professor, Cardozo School of Law, Yeshiva University: Crawford is also a policy analyst with the Center for Democracy & Technology and a Fellow with the Yale Law School Information Society Project. Her research and teaching interests include cyberlaw and intellectual-property law. While working as a partner at Wilmer, Cutler & Pickering (Washington, D.C.), she represented major online companies, start-ups, and joint ventures, working closely with companies doing business in the domain-name world. She is on the board of directors of Innovation Network, a nonprofit that helps other nonprofits develop and share evaluation tools and know-how.

Michael Dahan, Ben Gurion University of the Negev, Israel: Dahan is an Israeli-American political scientist living in Jerusalem and teaching at Ben Gurion University. His works include the paper "National Security and Democracy on the Internet in Israel." He has led projects to foster peace in the Middle East through new technology. One is an e-mail project that links political science students at Cairo University and Hebrew University. A second, more ambitious project is the Middle East Virtual Community of Academics and Intellectuals, which seeks to provide neutral ground for the exchange of ideas among intellectuals and to explore ways in which to break down the resistance to normalization with Israel.

Peter Denning, chairman, Naval Postgraduate School, Monterey, California: Denning is author of an IT column for *Communications of the ACM*. He has also been honored as an outstanding computer-science educator by the ACM. His area of research/teaching interest is the design of secure, reliable, dependable operating systems that meet performance requirements. His work is training officers to design and deploy information technology effectively.

Tobey Dichter, CEO, Generations on Line: Dichter founded this nonprofit Internet-literacy agency for seniors. She earlier worked as a vice president for public affairs at SmithKline Beecham.

Bill Eager, professional speaker: Eager is known for his presentations and workshops on business applications for the Internet. His is one of the voices in the Early 1990s Predictions Database. An Internet marketing pioneer, he wrote many books about the field, including the bestsellers *The Information Payoff* and *The Complete Idiot's Guide to Online Marketing*. While communications director at BASF, he was in charge of developing one of the first large-scale intranets in the United States.

Peter Excell, University of Bradford, U.K.: Excell is a professor of applied electromagnetics and director of research in the School of Informatics at the University of Bradford. He is also deputy director of UB's Telecommunications Research Centre.

Margot Edmunds, Johns Hopkins Department of Health Policy and Management: Edmunds is also former senior program officer at the Institute of Medicine in Washington, D.C., and director of health policy for the Children's Defense Fund.

Tom Egelhoff, smalltownmarketing.com: Egelhoff wrote a book about how small-town businesses can succeed in marketing and advertising, and he then put it on his Web site and continued sharing information. After 5 years, the site had grown to more than 300 pages of free tips and articles for small-business owners, and it had 4 million visitors in 2003. Egelhoff found a worldwide niche for advice on a small scale when he put his business on the Internet.

Ted Eytan, MD, Group Health Cooperative: Eytan is medical director of Group Health's online communications with patients—www.mygrouphealth.com.

Stan Felder, founder and president, Vibrance Associates: Felder's organization publishes the health/medical Web sites hisandherhealth.com, newshe.com, and ourgyn.com.

Howard Finberg, Poynter Insitute for Media Studies: Finberg, the director of Poynter's e-learning project—has been a Senior Fellow at the American Press Institute's Media Center, was a codirector of a year-long study on digital journalism for the Online News Association, and was named the Newspaper Association of America "New Media Pioneer" in 2000. A journalist for 30 years, he has worked at the *Chicago Tribune*, *The New York Times*, the *San Francisco Chronicle*, and the *Arizona Republic*. He is developing Poynter's News University, an online training portal for journalists.

Charles M. Firestone, The Aspen Institute: Firestone is executive director of the Communications & Society Program at Aspen, and has been there since 1989. The program focuses on the implications of communications and information technologies for leadership. He was previously director of the Communications Law Program at UCLA and president of the Los Angeles Board of Telecommunications Commissioners. The Aspen Institute offers seminars and special programs tailored to promote nonpartisan inquiry and leadership development.

Joshua Fouts, executive director of USC Center on Public Diplomacy: Fouts is a cofounder and the former editor of *Online Journalism Review*, based at USC. He previously spent 5 years at Voice of America, where he worked at getting VOA online with RealAudio.

Dan Froomkin, washingtonpost.com: Froomkin writes a political column for the online version of the *Washington Post* and is also the deputy editor of niemanwatchdog.org, based at Harvard University—a project to encourage more informed reporting by U.S. journalists. He also writes for the *Online Journalism Review*.

B. Keith Fulton, vice president, strategic alliances, Verizon Communications: Fulton was a senior telecommunications policy analyst before joining Verizon in 2004. He was a member of the U.S. Department of Commerce IPv6 Task Force, which examined issues associated with the next generation of the Internet protocol.

Simson L. Garfinkel, MIT; Sandstorm Enterprises; *Technology Review* magazine; *CSO* magazine: Garfinkel is one of the voices in the

Early 1990s Predictions Database. A journalist, entrepreneur, and international authority on computer security, he serves as chief technology officer at Sandstorm Enterprises, a Boston-based firm developing computer-security tools. He is a columnist for *Technology Review* magazine and has written tech articles for more than 50 publications, including *Computerworld*, *Forbes* and *The New York Times*. He is the author of *Database Nation*, *PGP: Pretty Good Privacy*, and many other books.

Christine Geith, Michigan State University: Geith is the director of product and business development at the Michigan State University Global Community Security Institute and was formerly executive director of e-learning at Rochester Institute of Technology, building one of the largest online learning programs in the U.S. She is on the executive committee of the board of directors for the National University Telecommunications Network.

Dan Gillmor, technology columnist, San Jose Mercury News: In addition to being a longtime tech writer, Gillmor is the author of *We the Media: Grassroots Journalism by the People, for the People*, a book about the participatory and citizen media movement. He also writes a popular Weblog.

Mark Glaser, *Online Journalism Review*: In addition to his writing about new media for *Online Journalism Review*, Glaser produces technology stories for the Online Publishers Association, CMP TechWeb, *The New York Times*' "Circuits" section, and *Condé Nast Traveler*.

Joshua Goodman, Microsoft Research: Goodman is a researcher in machine learning and language modeling at Microsoft who has been tasked to work on fighting spam. He also served as chairman of the second Conference on E-mail and Anti-Spam.

Moira K. Gunn, *Tech Nation*: Gunn was labeled the "grand dame of tech talk" in *Wired* magazine. She is the host of *Tech Nation*, National Public Radio's only syndicated technology talk show.

She also hosts technical Webcasts online for corporations such as Marimba and Network Associates, and she has conducted tech interviews packaged for public-television programming. She holds advanced degrees in computer science.

Alex Halavais, State University of New York at Buffalo: Halavais is graduate director for the informatics school. He studies how social networks are formed on the Internet. He promotes the practice of "self-Googling"—establishing your own identity on the Internet, so when people search out information about you it will be accurate.

Bornali Halder, Web site officer, World Development Movement: Halder works with an organization that lobbies decision makers to stop policies that hurt the world's poor. It researches and develops positive policy options that support sustainable development.

Fred Hapgood, Output Ltd.: Hapgood is an accomplished freelance writer in technology and science, and his is one of the voices in the Early 1990s Predictions Database. He took on the role of moderator of the Nanosystems Interest Group at MIT, and wrote a number of articles for *Wired* and other tech publications of the early 1990s.

Fran Hassencahl, professor, Old Dominion University: Hassencahl works in the Department of Communication at Old Dominion and is on the advisory board for InterculturalRelations.com. One of her specialties is Middle East affairs.

Brendan Hodgson, Hill & Knowlton: Hodgson is the director of Internet communications at Hill & Knowlton.

Robert Hughes Jr., University of Illinois-Champaign-Urbana: Hughes is an associate professor and extension specialist in the field of family relations. He is coauthor of the chapter "Understanding the Effects of the Internet on Families" in M. Coleman and L. H. Ganong (Eds.), *Handbook of Contemporary Families*, and author of the journal article, *Computers, the Internet and Families: A Review of the Role of New Technology in Family Life.*

Nigel Jackson, lecturer in public relations, University of Bournemouth, U.K.: Jackson worked as a staff member for a British political party and a member of Parliament, became a parliamentary lobbyist and consultant, led the communications departments of several organizations, and now teaches public relations. His research interests include the Internet and e-mail. He is on the editorial board of the *Journal of E-Government.*

Ken Jarboe, Athena Alliance: Jarboe is founder of a nonprofit, Washington, D.C.-based think tank that focuses on the social and economic implications of the Internet. Its aim is to change policy makers' thinking from the ways of industrial society, to those of the "networked society," by the means of conferences, workshops, research groups, publications, reports, and lectures. Areas of interest include digital empowerment and the digital divide. He is a former professor and University of Michigan policy/technology PhD.

Rich Jaroslovsky, senior editor, *Bloomberg News*: Jaroslovsky began his career as a *Wall Street Journal* reporter in 1975, served as White House correspondent and national political editor and became responsible for the *Wall Street Journal*/NBC News Polls in the 1980s. He created the *Wall Street Journal Interactive Edition* in 1994, and was executive director of editorial content for Dow Jones Consumer Electronic Publishing. He was the founding president of the Online Journalism Association.

Henry Jenkins, Massachusetts Institute of Technology: Jenkins is director of the Comparative Media Studies program at MIT, and a full professor of literature. He is the author of *Democracy and New Media* and coeditor of *Rethinking Media Change: The Aesthetics of Transition.*

Lyle Kantrovich, Cargill: Kantrovich is an Internet usability expert who works for the ag production company, Cargill. He is also a blogger who shares his thoughts on usability, Web design, information architecture, and user experience practices at *Croc o' Lyle*.

Daniel Kaplan, France's Next-Generation Internet Foundation (FING): Kaplan is the founder and CEO of FING, a collective and open project focusing on future Internet uses, applications, and services. He is also chairman of the European Institute for e-Learning (EifEL). He is a member of the European Commission's e-Europe's Experts Chamber, the French Prime Minister's Strategic Advisory Board on Information Technologies (CSTI), and the board of the French Chapter of the Internet Society.

Ruth Kaufman, IBM: Kaufman is a Web strategist based in IBM's offices in White Plains, New York.

Mike Kelly, America Online: Kelly is a leading executive for one of the nation's most-recognizable Internet "brands."

Yonnie Kim, chief researcher, Daum Communication Thinktank: Kim specializes in new-trend research and strategy development. Daum operates South Korea's largest Internet portal.

Gary Kreps, chair, Department of Communication, George Mason University: Kreps also holds a joint faculty appointment with the National Center for Biodefense at GMU. Before his appointment at GMU, he served for 5 years as the founding chief of the Health Communication and Informatics Research Branch at the National Cancer Institute.

Anne Laurent, associate editor, *Government Executive Magazine*: Laurent specializes in stories about entrepreneurial organizations, acquisition reform, results-based management, and culture change, and she manages the Government Performance Project.

Douglas Levin, Cable in the Classroom: Levin is education-policy director for the group that represents the cable television industry's commitment to education. He previously served as a principal research analyst with the American Institutes for Research in

Washington, D.C., where he wrote national studies on the role of technology in education, including "The Digital Disconnect," conducted on behalf of the Pew Internet & American Life Project. He also assisted in the development of three U.S. Department of Education National Education Technology plans.

Peter Levine, University of Maryland: Levine is a research scholar at the Institute for Philosophy & Public Policy and deputy director of the Center for Information and Research on Civic Learning and Engagement, both at the University of Maryland's School of Public Policy. He formerly worked for Common Cause. His main interests are civil society, civic education, and the Internet. He also works with the Prince George's County Information Commons (a nonprofit Web site for the community, produced mainly by youth), The National Alliance for Civic Education (as the person responsible for the Web site), and the Deliberative Democracy Consortium.

Graham Lovelace, managing director, Lovelacemedia Ltd., U.K.: Lovelace is founding director of his company, which he started after leaving the *Daily Mail Group*, where he was editorial director at Teletext Limited and director of Associated New Media. In the 1990s, he was a pioneer in Internet publishing and digital broadcasting in the U.K. He worked as a senior editor and journalist at the BBC, Visnews (now Reuters TV), BSB (now BSkyB), ITN, Channel Four, the Press Association, and Pearson regional Press. He is a regular commentator on new media in newspaper columns and in television and radio interviews.

Robert Lunn, FocalPoint analytics: Lunn worked as a senior research analyst on the 2004 Digital Future Report: Surveying the Digital Future, produced by the USC Annenberg School Center for the Digital Future.

Shawn McIntosh, Columbia University/Netgraf: McIntosh is coauthor of *Converging Media* and worked as an editor and freelance writer for newspapers and magazines in the U.K., U.S., and Japan before working at Fathom, an online educational Web site made up of a consortium of academic institutions, museums, and research organizations. He cofounded Netgraf, which examines issues and trends related to online journalism.

John B. Mahaffie, cofounder, Leading Futurists LLC: Mahaffie is a former principal of Coates & Jarratt Inc., a leading futures consultancy. He has been a futurist since the mid-1980s. Clients currently include the Coca-Cola Company, DuPont, General Motors, Nokia, Siemens, the National Security Agency, and the World Bank. He is also a board member of the Association of Professional Futurists.

Michelle Manafy, Information Today, Inc.: Manafy is editor of *EContent* magazine and *Intranets* newsletter. She is a former associate editor of *EMedia* magazine. She has written about content development and distribution, streaming media, and audio, video, and storage technologies.

Vikram Mangalmurti, Carnegie Mellon University: Mangalmurti works at the H. John Heinz School of Public Policy and Management. His research interests include security and privacy.

J. Scott Marcus, senior adviser for Internet technology at the Federal Communications Commission: Marcus previously served as chief technology officer at Genuity (GTE Internetworking) and does research in the economics and public-policy implications of network interconnection—backbone connections in particular. He specializes in the measurement and prediction of Internet usage, challenges of data network security, and management of data networks. He served as a trustee of the American Registry of Internet Numbers from 2000 to 2002.

Jon Marshall, Research Fellow, University of Technology, Sydney: Marshall has done research on gender in online interaction and anethnography of the mailing list *Cybermind*; his most recent publications include articles on the construction of the Internet as "space" and on netsex in an online community.

Bob Metcalfe, Polaris Venture Partners: Metcalfe is a venture capitalist—an early stage investor in bio-, info- and nano-technology companies. He is the inventor of Ethernet and founded 3Com Corporation, the billion-dollar networking company. He was CEO of InfoWorld Publishing from 1992 to 1995 and wrote a popular column for information professionals for 8 years. His is one of the voices in the Early 1990s Predictions Database. His books include *Packet Communication*, *Beyond Calculation: The Next Fifty Years of Computing*, and *Internet Collapses*.

Ezra Miller, Ibex Consulting, Ottawa, Canada: Miller's fields of concentration in the consulting arena are e-government, ICT policy, and economics.

Kirsten Mogensen, associate professor, Roskilde University, Denmark: Mogensen teaches in the Department of Journalism at Roskilde University.

Arlene Morgan, Columbia Graduate School of Journalism: Morgan, formerly assistant managing editor of the *Philadelphia Inquirer*, is director of professional development at Columbia's Graduate School of Journalism.

Dan Ness, MetaFacts: Ness is principal consultant at MetaFacts, a market-research firm that solves customer challenges for high-tech companies such as Advanced Micro Devices, Adobe, Compaq, Dell, Gateway, Hewlett Packard, IBM, Microsoft, MTV, Sony, and Toshiba. He is a 20-plus-year veteran of the computer industry with extensive experience in both primary and secondary research. During the last 2 decades, his research projects and programs have included more than 1.5 million interviews, including surveys of businesses and consumers, both domestically and internationally. He has participated in product design, launches, repositioning, branding, and pricing with most of the leading high-tech worldwide companies.

Biographies

Mike O'Brien, The Aerospace Corporation: O'Brien founded and ran the first nationwide UNIX Users Group Software Distribution Center. He worked at RAND and helped build CSNET (first at RAND and later at BBN Labs Inc.). He now works at an aerospace research corporation.

Ahmet M. Oren, chairman, Ihlas Holding, Turkey: Oren, a leader at Ihlas, a major Middle East television and news-gathering organization based in Istanbul, has been a member of the board of directors for the International TV Academy and Interactive Television International.

Andy Opel, Dept. of Communication, Florida State University: Opel has written about micro radio and the emerging media activism movement. He edited the book, *Representing Resistance: Media, Civil Disobedience and the Global Justice Movement*, published by Greenwood Press.

George Otte, City University of New York: Otte is on the doctoral faculty of the Graduate Center Programs in English, Urban Ed, and Technology & Pedagogy and is director of instructional technology at CUNY.

Carlos Andrés Peña, Novartis Pharma: Peña is an expert on evolutionary computation in medicine.

Terry Pittman, executive director broadband markets, America Online: Pittman heads the broadband division of AOL. He is active as a member of TRUSTe's board of directors—TRUSTe is a leading online privacy seal program. Pittman also founded Postmodern Media, a new-media consulting firm; was advertising director for Netcom; and worked with BrightStreet.com. He is an expert on online privacy and has been an active participant in the privacy discussion since the mid-'90s.

Louis Pouzin, Eurolinc France: Pouzin conceived and directed the Cyclades project at the Institut de Recherche d'Informatique et d'Automatique in France. This project laid the conceptual groundwork Vinton Cerf and others employed in building the Internet. Before that, Pouzin spent time at MIT and also worked for several large companies, including Chrysler, mainly in developing advanced operating systems. In 2001 Pouzin received the Institute of Electrical and Electronic Engineers (IEEE) Internet Award for his work in datagram networking.

Sam Punnett, president, FAD Research, Toronto, Canada: Punnett has worked in the field of interactive digital media since the 1980s. He has worked in the music business, social research, broadcast production, equities analysis, electronic game design, and for the last 9 years on strategy, marketing, and product development issues related to e-business. He has published numerous studies and written extensive commentary on new technology for private companies and government agencies interested in Internet economy issues.

Brian Reich, strategic consultant, Mindshare Interactive Campaigns: Reich develops online strategies for issue coalitions and nonprofit organizations. He is a former national Democratic political operative and Internet strategist. He was Vice President Al Gore's briefing director in the White House and during Gore's 2000 presidential campaign. Reich also has a consulting firm, Mouse Communications, and is the editor of the political blog, *Campaign Web Review* (www.campaignwebreview.com).

Paul Resnick, University of Michigan: Resnick is conducting research with Bob Kraut, Sara Kiesler, Yan Chen, Loren Terveen, John Riedl, and Joe Konstan on the public good in online communities; the project is being funded by the National Science Foundation. He has also worked in the areas of online reputation systems and in studying Meetup.com and other "convening technologies."

Howard Rheingold: Rheingold was one of the first writers to illuminate the ideals and foibles of virtual communities. In the 1990s, he published a Webzine called *Electric Minds*. He wrote the books *Virtual Reality*, *Smart Mobs*, and *Virtual Community*. He also was the editor of *Whole Earth Review* and the *Millennium Whole Earth Catalog*. He is a popular commentator on the human-to-human implications of the Internet.

Victor Rivero, editor, writer, consultant: A former editor of *Converge* educational technology magazine, Rivero is a journalist specializing in education technology.

Mark Rovner, CTSG/Kintera: Rovner's work at CTSG is online fundraising, engagement, and communications strategizing. He has spent his 20-year career working with fundraising projects in the nonprofit and political sectors. He previously served as a senior vice president at Craver, Mathews, Smith & Co. (CMS) where he was founding director of CMS Interactive, the firm's Internet fundraising unit. At CMS, he oversaw online fundraising strategy for a number of the nation's leading charities and advocacy groups, including Amnesty International and the American Civil Liberties Union.

Douglas Rushkoff, author and a teacher in New York University's Interactive Telecommunications Program: Rushkoff is one of the voices in the Early 1990s Predictions Database. The successful author is also a teacher in the New York University Interactive Telecommunications Program. This social theorist, journalist, and software developer wrote the book *Cyberia: Life in the Trenches of Hyperspace*, a best-selling portrait of the 1990s cyberculture. He edited *The Gen X Reader*, a collection of writings by the elusive, media-wary "slacker" generation. He also wrote *Media Virus! Hidden Agendas in Popular Culture*, *Exit Strategy*, and *Coercion*, and he was a winner of the Neil Postman Award for Career Achievement in Public Intellectual Activity. He is also the author of *Open Source Democracy*, written for the U.K. policy think tank Demos.

Cheryl Russell, New Strategist Publications: Russell has been labeled the "patron saint of Boomer research." She is the author of

The Baby Boom: Americans Born 1946 to 1964 and is the editorial director of New Strategist Publications. Her other books include *The Official Guide to the American Marketplace*, and *Demographics of the U.S.: Trends of Projections*.

Liz Rykert, Meta Strategies Inc.: Rykert is president and founder of a Canadian consulting firm that develops digital strategies for governments. She drafted guidelines for the Canadian government to steer online consultation and engagement work for federal departments. She also completed a project for a group in Malawi to track AIDS/HIV work in Sub-Saharan Africa. One of her research/consulting areas has been digital democracy and advocacy-based citizen engagement.

Janet Salmons, Vision2Lead.inc: Salmons consults in organizational and leadership development. A new initiative, Elearn2Lead, focuses on leadership development and online learning. Current projects involve work with the Lynn University Institute for Distance Learning, Southeastern Community College Distance and E-Learning, the National Endowment for the Arts Theatre Program, and TechSoup. She is a frequent conference speaker—on and offline. She presented Virtual Learning Community: Training Staff & Volunteers for the Wired.org: Nonprofits and NGOs Work the Web online conference, and Online Communities to Enhance Learning for Your Organization, a one-week online event.

Bill Sanders, senior vice president, Paramount Television: Sanders was a cofounder of Big Ticket Television, a Paramount/Viacom company and developed its first eight on-air series. He developed an enhanced interactive version of *Judge Judy* for Web TV and also produced broadband trials for *Judge Joe Brown*. He developed one of the first TV-show-based Web sites—including online merchandising—in 1994. He worked as a vice president for West Coast programming for HBO, where he was supervising producer of the multiple Emmy-winning series, *Dream On*, for which he developed online chat forums and an interactive CD-ROM game.

Alexandra Samuel, research and communications director, Harvard University and Cairns Project (New York Law School): The Cairns Project is building civic software to promote problem solving and decision making to engage citizens in active democracy. Her areas of study/research interest include politically motivated computer hacking, e-government, and e-business strategy. Samuel worked as the research director for Governance in the Digital Economy, an international research program.

Daniel Z. Sands, Zix Corporation, Beth Israel Deaconess Medical Center and Harvard Medical School: Sands is an internationally recognized lecturer, consultant, and thought leader in the area of clinical computing and patient and clinician empowerment through the use of computer technology. He received the President's Award from the American Medical Informatics Association for coauthoring the first national guidelines for the use of e-mail in patient care. In 2003 he was elected to the American College of Medical Informatics and was granted an IT Innovator award by *Healthcare Informatics* magazine for his leadership in advancing electronic patient-centered communication.

Jan Schaffer, executive director, J-Lab: Schaffer runs J-Lab: The Institute for Interactive Journalism, based at the University of Maryland. She was a Pulitzer-winning reporter at the *Philadelphia Inquirer* and was executive director of the Pew Civic Journalism Project before founding J-Lab at the University of Maryland.

Jorge Reina Schement, Penn State University: He is director of the Institute for Information Policy at Penn State. His book credits include *Tendencies and Tensions of the Information Age*, *Toward an Information Bill of Rights and Responsibilities*, and *The Wired Castle*, a study of information technology in American households. His research interests focus on the social and policy consequences of the production and consumption of information. In 1994 he served as director of the FCC's Information Policy Project. A member of the boards of directors of the Media Access Project, Libraries for the

Future, and the Benton Foundation, he regularly leads seminars at The Aspen Institute.

David M. Scott, communications strategist, Freshspot Marketing: Scott founded Freshspot, which serves information-product and IT-service companies. He is a contributing editor for *EContent* magazine, a source for strategies and resources for the digital content industry.

Tiffany Shlain, founder and chairperson, the Webby Awards: Shlain was named one of Newsweek's "Women Shaping the 21st Century." The Webby Awards are the leading international honors for Web sites. Shlain has also directed 10 films, including *Life, Liberty and the Pursuit of Happiness*, an official selection at the 2003 Sundance Film Festival, and *Less is Moore*, a profile of Intel founder Gordon Moore. She is also an expert commentator on Internet issues, appearing regularly on television and radio. She is a Fellow of the Woodhull Institute, an organization that supports ethical women leaders for the next century.

Barbara Smith, Institute of Museum and Library Services: Smith is technology officer for a federal agency that offers support to all types of museums, libraries, and archives.

Robert V. Steiner, American Museum of Natural History: Steiner is the project director for seminars on science at the American Museum of Natural History. It offers online courses to K–12 teachers across the United States in the life, earth, and physical sciences, as well as courses about broader trends. He also works with the Teachers College of Columbia University in the Center for the Study of the Science of Learning in Urban Educating Institutions.

William Stewart, LivingInternet.com: Stewart has worked as a program manager, system architect, system engineer, software engineering manager, software lead, computing center manager, and university instructor. But the thing he is known for is LivingInternet.

com, a site first developed in 1996 and devoted to explaining and using the Internet. Many Internet pioneers have made contributions to the more than 500 pages on the site.

Gordon Strause, Judy's Book: Strause works for Judy's Book, a social-networking Web site founded in the Seattle area in 2004. It allows friends and coworkers to post personal reviews of mechanics, dentists, house painters, and so forth that they would like to share with their online community. He formerly worked at Firefly, Well Engaged, and eCircles.

Kevin Taglang, consultant and telecommunications policy expert: Taglang's clients include the Benton Foundation, whose mission is to articulate a public-interest vision for the digital age and to demonstrate the value of communications for solving social problems. He has published research on the digital divide.

Peter W. Van Ness, principal, the Van Ness Group: The Van Ness Group, a Web development company, is the third technology-related venture for entrepreneur Peter Van Ness. He founded Personal Computer Solutions in 1983. During the 1990s, he cofounded StockPlan, Inc., and grew it from a tiny software startup into the largest independent provider of stock-plan management services worldwide; there, he and his team built the first systems for employees to exercise and sell their stock options over the Internet. In 2000 he cofounded MyStockOptions.com, winner of numerous awards and the Web's most comprehensive, respected, and frequently visited resource on stock compensation.

Egon Verharen, innovation manager, SURFnet: Verharen works for the Dutch national education and research network. He was previously an assistant professor of information technology at Infolab at Tilburg University.

Rose Vines, freelance technology journalist: Vines writes for *Australian PC User* and the *Sydney Morning Herald*.

Philip Virgo, EURIM and IMIS: Virgo is secretary general of EURIM, the U.K.-based Parliament-Industry group, and he is also associated with IMIS, a U.K.-based professional body for management of information systems. He wrote a research study that projected "The World and Business Computing in 2051."

Barry Wellman, University of Toronto: Wellman's research examines virtual community, the virtual workplace, social support, community, kinship, friendship, and social network theory and methods. He directs NetLab and does research at the Centre for Urban and Community Studies, the Knowledge Media Design Institute, and the Bell University Laboratories' Collaborative Effectiveness Lab. He has been a Fellow of IBM's Institute of Knowledge Management, a member of Advanced Micro Devices' Global Consumer Advisory Board, and a committee member of the Social Science Research Council's Program on Information Technology, International Cooperation and Global Security. He is the author or coauthor of more than 200 articles, coauthored with more than 80 scholars, and is the editor or coeditor of three books. He is conducting a "Strong Ties and Weak Ties On and Offline" study for the Pew Internet & American Life Project.

David Weinberger, Evident Marketing Inc.: Weinberger is a writer, speaker, and consultant on Internet communication, publishing, and marketing, and he a columnist for MIT's *Technology Review, Darwin Online, Intranet Desig,* and *Knowledge Management World* and has been a frequent commentator on National Public Radio's *All Things Considered* and *Here and Now*. He is one of the voices in the Early 1990s Predictions Database. His one-person consulting company has served a wide range of IT clients, including Sun Microsystems and Esther Dyson's Release 1.0. He is a coauthor of *The Cluetrain Manifesto* and author of *Small Pieces Loosely Joined*, both about the Internet. He is a Fellow at Harvard's Berkman Institute for Internet and Society. He writes several Weblogs and was senior Internet adviser for Howard Dean's 2004 presidential campaign.

Mike Weisman, attorney based in Seattle: Weisman, an active leader of the advocacy group, Reclaim the Media, and a community technology activist in the Seattle area, represented the Center for Digital Democracy and Consumer Federation of America in hearings regarding the AT&T Comcast merger. He was the program director for the 2003 conference of the Computer Professionals for Social Responsibility.

Pamela Whitten, Michigan State University: Whitten is an associate professor in the Department of Telecommunications at Michigan State and is a Senior Research Fellow at its Institute of Healthcare Studies. Her research focuses on the use of technology in health care, and her research projects range from telepsychiatry to telehospice and telehome care for COPD and CHF patients. She serves on the board of directors of the American Telemedicine Association.

Mike Willard, chief executive officer of the Willard Group: Willard heads Burson-Marstellar's affiliate in Eastern Europe. He has been a newsman, political and policy advisor to U.S. senators, senior public relations counselor, and entrepreneur. He is a specialist in crisis communications and management, and served as a communications and domestic and foreign policy advisor to U.S. Senator Robert C. Byrd and as a consultant to Senator John D. Rockefeller, when he was a governor.

Roland B. Wilson, Indiana State University: Wilson works in IU's Languages, Literatures & Linguistics Lab.

Michael Wollowski, professor, Rose-Hulman Institute of Technology: Wollowski is an assistant professor of computer science and software engineering.

Steve Yelvington, Morris Digital Works/Morris Communications: Yelvington put the *Minneapolis Star Tribune* online in 1994. He became executive editor of Cox Interactive Media in 1999, and now he is vice president of strategy and content for the interactive division of Morris Newspapers, whose newspaper Web sites have won more awards than any other newspaper chain in America. In 2001 the Newspaper Association of America presented him with its New Media Pioneer Award.

Appendix III

Survey Results and Methodology

Future of the Internet I Final Combined Topline 12/06/04
Web Survey

Data for September 20–November 1, 2004

Princeton Survey Research Associates for the Pew Internet & American Life Project

Sample: *n* = 1,286 Internet experts from Wave 1 and Wave 2 Interviewing dates: 09.20.04–11.01.04

Note. The current survey results are based on a nonrandom online sample of 1,286 Internet stakeholders, recruited via e-mail invitation from the Pew Internet & American Life Project. Since the data are based on a nonrandom sample, a margin of error cannot be computed, and the results are not projectable to any population other than the experts in this sample.

Q1 How did you learn of this survey?

	Current	
%	81	E-mail invitation from the Pew Internet & American Life Project
	15	E-mail alert from another person
	1	Posting on a Web site
	2	Another way
	1	Did not respond

Q2 What year did you first start using the Internet?

	Current	
%	6	1982 or earlier
	38	Between 1983 and 1992
	54	1993 or later
	1	Did not respond

Q3 What is your primary area of Internet interest?

	Current	
%	17	Research Scientist
	14	Entrepreneur/Business Leader
	13	Author/Editor/Journalist
	11	Technology Developer/Administrator
	8	Futurist/Consultant
	6	Advocate/Voice of the People/Activist User
	2	Legislator/Politician/Lawyer
	1	Pioneer/Originator
	27	Other
	1	Did not respond

Q4 What is the name of the organization where you work? If you currently work for more than one organization, please feel free to list them all.

Q5 What type of organization is that? If you work for more than one organization, circle all that apply.

	Current	
%	31	A college or university
	16	A nonprofit organization
	15	A company whose focus is not mainly on information technology but extensively uses it
	13	A company whose main focus is on information technology
	13	A consulting business

	Current (continued)
11	A government agency
8	A research organization
5	A publication
15	Other
2	Did not respond

Note. Table total exceeds 100% due to multiple responses.

Q5a If you would like, please give us your name.

Q5b If you would like, please give us your title.

Q5c If you were forwarded an e-mail invitation from another person, please tell us how you found out about this survey.

Q6 On a scale of 1–10 with 1 representing no change and 10 representing radical change, please indicate how much change you think the Internet will bring to the following institutions or activities <u>in the next decade</u>.

	1	2	3	4	5	6	7	8	9	10	Did Not Resp.	Mean
a. Politics and government	*	1	3	5	7	11	19	22	14	16	2	7.39
b. Education	*	*	1	3	5	8	16	21	19	24	2	7.98
c. Medicine and health care	*	1	3	4	7	10	15	20	17	20	2	7.63
d. Workplaces	*	*	1	2	7	9	18	22	18	20	2	7.84
e. Military	1	3	7	7	14	11	16	15	10	12	4	6.53
f. News organizations and publishing	*	*	1	1	3	5	12	20	23	33	2	8.46
g. Music, literature, drama, film, and the arts	*	2	4	7	10	10	15	19	14	16	2	7.18
h. Families	*	3	8	9	17	15	16	16	7	7	2	6.24
i. Neighborhoods and communities	1	5	7	9	17	14	19	14	6	7	2	6.16
j. Religion	5	14	16	11	16	11	11	7	2	3	3	4.69
k. International relations	1	3	7	5	13	12	17	17	9	13	2	6.74

Q7 **Elaborate your thinking:** In the next decade, which institutions and human endeavors will change the most because of the Internet? Tell us how you see the future unfolding or point us to your favorite recent statement about the impact of the Internet in the future.

Q8a **Prediction on Social Networks.**
By 2014, use of the Internet will increase the size of people's social networks far beyond what has traditionally been the case. This will enhance trust in society, as people have a wider range of sources from which to discover and verify information about job opportunities, personal services, common interests, and products.

	Current	
%	39	Agree
	20	Disagree
	27	Challenge the prediction
	15	Did not respond

Q8b **Please explain your answer.** If you like, please elaborate further.

Q9a **Prediction on Attacks on Network Infrastructure.**
At least one devastating attack will occur in the next 10 years on the networked information infrastructure or the country's power grid.

	Current	
%	66	Agree
	11	Disagree
	7	Challenge the prediction
	16	Did not respond

Q9b **Please explain your answer.** If you like, please elaborate further.

Q10a Prediction on Digital Products.
In 2014 it will still be the case that the vast majority of Internet users will easily be able to copy and distribute digital products freely through anonymous peer-to-peer networks.

	Current	
%	50	Agree
	23	Disagree
	10	Challenge the prediction
	17	Did not respond

Q10b Please explain your answer. If you like, please elaborate further.

Q11a Prediction on Civic Engagement.
Civic involvement will increase substantially in the next 10 years, thanks to ever-growing use of the Internet. That would include membership in groups of all kinds, including professional, social, sports, political, and religious organizations—and perhaps even bowling leagues.

	Current	
%	42	Agree
	29	Disagree
	13	Challenge the prediction
	17	Did not respond

Q11b Please explain your answer. If you like, please elaborate further.

Q12a Prediction on Embedded Networks.
As computing devices become embedded in everything from clothes to appliances to cars to phones, these networked devices will allow greater surveillance by governments and businesses. By 2014, there will be increasing numbers of arrests based on this kind of surveillance by democratic governments, as well as by authoritarian regimes.

	Current	
%	59	Agree
	15	Disagree
	8	Challenge the prediction
	17	Did not respond

Q12b Please explain your answer. If you like, please elaborate further.

Q13a Prediction on Formal Education.
Enabled by information technologies, the pace of learning in the next decade will increasingly be set by student choices. In 10 years, most students will spend at least part of their "school days" in virtual classes, grouped online with others who share their interests, mastery, and skills.

	Current	
%	57	Agree
	18	Disagree
	9	Challenge the prediction
	17	Did not respond

Q13b Please explain your answer. If you like, please elaborate further.

Q14a Prediction on Democratic Processes.
By 2014, network security concerns will be solved and more than half of American votes will be cast online, resulting in increased voter turnout.

	Current	
%	32	Agree
	35	Disagree
	15	Challenge the prediction
	18	Did not respond

Q14b Please explain your answer. If you like, please elaborate further.

Q15a Prediction on Families.
By 2014, as telework and homeschooling expand, the boundaries between work and leisure will diminish significantly. This will sharply alter everyday family dynamics.

	Current	
%	56	Agree
	17	Disagree
	9	Challenge the prediction
	18	Did not respond

Q15b Please explain your answer. If you like, please elaborate further.

Q16a Prediction on the Rise of Extreme Communities.
Groups of zealots in politics, in religion, and in groups advocating violence will solidify, and their numbers will increase by 2014 as tight personal networks flourish online.

	Current	
%	48	Agree
	22	Disagree
	11	Challenge the prediction
	19	Did not respond

Q16b Please explain your answer. If you like, please elaborate further.

Q17a Prediction on Politics.
By 2014, most people will use the Internet in a way that filters out information that challenges their viewpoints on political and social issues. This will further polarize political discourse and make it difficult or impossible to develop meaningful consensus on public problems.

	Current	
%	32	Agree
	37	Disagree
	13	Challenge the prediction
	18	Did not respond

Q17b Please explain your answer. If you like, please elaborate further.

Q18a Prediction on Health System Change.
In 10 years, the increasing use of online medical resources will yield substantial improvement in many of the pervasive problems now facing health care—including rising health care costs, poor customer service, the high prevalence of medical mistakes, malpractice concerns, and lack of access to medical care for many Americans.

	Current	
%	39	Agree
	30	Disagree
	11	Challenge the prediction
	19	Did not respond

Q18b Please explain your answer. If you like, please elaborate further.

Q19a Prediction on the Personal Entertainment and Media Environment.
By 2014, all media, including audio, video, print, and voice will stream in and out of the home or office via the Internet. Computers that coordinate and control video games, audio, and video will become the centerpiece of the living room and will link to networked devices around the household, replacing the television's central place in the home.

	Current	
%	53	Agree
	18	Disagree
	10	Challenge the prediction
	19	Did not respond

Q19b Please explain your answer. If you like, please elaborate further.

Q20a Prediction on Creativity.
Pervasive high-speed information networks will usher in an Age of Creativity in which people use the Internet to collaborate with others and take advantage of digital libraries to make more music, art, and literature. A large body of independently produced creative works will be freely circulated online and will command widespread attention from the public.

	Current	
%	54	Agree
	18	Disagree
	9	Challenge the prediction
	20	Did not respond

Q20b Please explain your answer. If you like, please elaborate further.

Q21a Prediction about How People Go Online.
By 2014, 90% of all Americans will go online from home via high-speed networks that are dramatically faster than today's high-speed networks.

	Current	
%	52	Agree
	20	Disagree
	8	Challenge the prediction
	20	Did not respond

Q21b Please explain your answer. If you like, please elaborate further.

Q22 Thinking back to your views a decade ago, where has the use or impact of the Internet fallen short of your expectations?

Q23 What impacts have been felt more quickly than you expected?

Q24 What are you anxious to see happen? What is your dream application, or where would you hope to see the most path-breaking developments in the next decade?

METHODOLOGY

Internet Expert Web Survey
Prepared by Princeton Survey Research Associates International for the Pew Internet & American Life Project

December 2004

SUMMARY

The Internet Expert Web Survey, sponsored by the Pew Internet & American Life Project, obtained online interviews with a nonrandom sample of 1,286 Internet experts. The interviews were conducted online, via SPSS, in two waves: Wave 1 took place from from September 20 to October 18, 2004, and Wave 2 took place from October 19 to November 1, 2004. Details on the design, execution, and analysis of the survey are discussed next.

Sample Design/Contact Procedures
Across both waves of the project, e-mail invitations to participate in the survey were sent to slightly less than 1,000 Internet experts (367 of these were sent *after* the completion of Wave 1 of the project). Overall, approximately 7% of the e-mail addresses proved invalid, for a working rate of 93%. The e-mail invitations provided a direct link to the survey, and contained the following language:

PEW/INTERNET
PEW INTERNET & AMERICAN LIFE PROJECT

Dear [name here]:

The Pew Internet & American Life Project is surveying experts about the future of the Internet and we would very much like to include your views in our research.

The idea for this project grew out of work we did with Elon University to develop a database of over 4,000 predictions about the impact of the Internet made by experts during the period between 1990 and 1995. Now we are conducting a Web-based survey about the impact the Internet might have in the next decade. We are canvassing many of the people whose predictions are included in the original 1990–1995 database—and we are soliciting predictions from other experts who have established themselves in recent years as thoughtful analysts.

We hope you'll take 10–15 minutes to fill out our survey, which you will find at http://surveys.spss-sb.com/spssmr/survey/surveyentry.aspx?project=p3280003. The survey asks you to assess several predictions about the future impact of the Internet and to contribute your own thoughts about what will happen in the next 10 years.

This is a confidential survey. However, we encourage you to take credit for your thoughts. When you start the survey, please use this personal identification number (PIN):

[PIN]

The Pew Internet & American Life Project will issue a report based on this survey during autumn; we hope the results will be useful to policy makers, scholars, and those in the information technology industry. Our goal is to include material from this new survey in the predictions database. (While we have not publicly talked about that effort yet, you can browse through the existing

> material at http://www.elon.edu/predictions.) Be assured that we will not use your name or e-mail address for any purpose other than this research project, and will not share your information with outside solicitors.
>
> We're sure we have not identified all experts whose views would be helpful to this research, so I would invite you to send an invitation to participate in this survey to any friends or colleagues whose insights you would be interested in learning. Please ask them to use **PIN 700** when taking the survey.
>
> I hope you enjoy taking the survey and sharing your views about the future of the Internet. If you have any questions, please feel free to contact me at lrainie@pewInternet.org.
>
> Thank you,
> Lee Rainie
>
> Director
> Pew Internet & American Life Project
> 1100 Connecticut Ave. NW
> Suite 710
> Washington, D.C. 20036
> 202.557.3463

As the aforementioned text indicates, Pew Internet encouraged the initial sample of experts to forward the e-mail invitation to any colleagues whose thoughts on the future of the Internet they would consider useful and important. This created an additional snowball sample of Internet experts, whose ideas are also included in the final data.

Completion Rate
Based on figures supplied by SPSS, PSRAI has calculated the following completion rate for the Experts Survey:

In Table 2, total hits (1,892) indicate the number of times the survey link was accessed between September 20 and November 1, 2004, or roughly the number of potential respondents who reached

Survey Results and Methodology

TABLE 2. Overall survey completion rate.

	Number	*Rate*
Total Hits	1,892	
Total Completes	1,286	68.0%
Final Completion Rate		*68.0%*

the survey's title page during the field period. The survey title page gave the following brief description of the survey and its sponsors, along with instructions for how to complete the survey:

Forecasting the Internet

Welcome to the Pew Internet & American Life Project survey of technology experts and social analysts about the future of the Internet. This survey asks you to assess some predictions and contribute your own thoughts about the impact of the Internet in the next 10 years.

This survey has grown out of as yet unpublished research by the Project and Elon University to study predictions made between 1990 and 1995 about the evolution of the Internet. The "Imagining the Internet" database of those predictions is available at http://www.elon.edu/predictions. We plan to update the database to include responses from this survey, as well as your unfiltered answers.

The project's goal is to see where experts agree and disagree about the potential social impact of the Internet. We hope the findings will illuminate issues for policy makers, spark debate and further research among scholars, and encourage those who build technology to ponder the societal effects of their creations.

This is a confidential survey. However, we encourage you to take credit for your thoughts. Please feel free to put your name in any space that allows for written answers. We will only credit to you the

> individual statements to which you add your name in the answer block. If your name is not there, your comments will be attributed to an anonymous voice when they are added to the Pew Predictions Database.
>
> We plan to publish the results of this survey in a report that will be issued this autumn.
>
> S1. If you received an e-mail invitation from Pew Internet with an individual PIN for taking this survey, please enter it here.
>
> Those who were invited to participate by a friend or colleague should use guest PIN 700.
>
> If you did not receive either an individual or guest PIN, please enter 999 and proceed.

Total completes (1,286) indicate the number of respondents who completed the survey through at least Question 6. The final completion rate for the survey is computed as the number of completes (1,286)/the number of hits (1,892), or 68.0%.

Questionnaire Development

The questionnaire was developed by PSRAI in collaboration with staff of the Pew Internet & American Life Project and their partners at Elon University.

Note About Survey Data

PSRAI recommends including the following note with any public release of the data:

Note. Results are based on a nonrandom online sample of 1,286 Internet stakeholders, recruited via e-mail notices sent to an initial sample of pre-identified experts, as well as a snowball sample of their colleagues. Since the data are based on a nonrandom sample, a margin of error cannot be computed, and the results are not projectable to any population other than those experts who completed the survey.

CONTRIBUTOR INDEX

Ambash, Lois, 23, 58, 89, 106, 122, 140, 167, 176, 193, 233, 244, 261, 275
Arlen, Gary, 21, 43, 88, 126, 208, 233–234, 282
Ashe, Reid, 55
Atkinson, Rob, 11, 109, 223, 249

Bachula, Gary, 120, 152, 191, 204, 244, 276
Baker, Paul M. A., 128, 167
Band, Jonathan, 37, 73, 107, 124, 152, 219, 276
Barrat i Esteve, Jordi, 19, 140
Blanchard, Anita, 92
Boase, Jeffrey, 279
Boese, Christine, 25, 45, 91, 128
Booher, Bill, 121
Botein, Mike, 10, 40, 122, 192, 207, 233, 243, 260, 276
Bracey, Bonnie, 94
Brancheau, James, 6, 204, 243
Breeden, Laura, 129, 235, 284
Breindahl, Charlie, 60
Brier, Steven, 235
Brown, Bradford, 7, 40, 86, 107, 125, 190, 205, 222, 247, 263, 281
Buerger, Michael, 60
Buys, Jay, 75

Carruthers, Kate, 56
Castro, Janice, 8, 248, 264
Chabot, Elliot, 75, 177
Chapman, Gary, 19, 39, 104, 123, 190, 222, 243, 276

Chodorow, Stanley, 41, 59, 70, 89, 106, 126, 192, 220, 281
Christensen, Ted, 12, 181, 250
Claburn, Thomas, 23
Coelho, Antonio, 74,
Compaine, Benjamin, 36, 105, 123, 165, 176, 190, 206, 221, 223
Contractor, Noshir, 37, 164, 174, 244
Coopman, Ted, 110, 142
Coppins, Steve, 57, 166, 234
Crawford, Joe, 244, 247
Crawford, Susan, 40, 72, 86, 104, 121, 141, 153, 164, 175, 220, 233

Dahan, Michael, 14, 91, 127, 283
DeCleene, Clare, 152
Denning, Peter, 20, 39, 74, 86, 121, 139, 191, 244, 260, 277
DeSantis, Joe, 244
DeVries, Richard, 110, 129
Dichter, Tobey, 85, 193, 205, 248
Douglas, Rushkoff, 20, 38, 73, 87, 105, 138, 151, 218, 243, 260
Drucker, Peter, 86, 162

Eager, Bill, 20, 176, 260, 277
Eccles, Bill, 278
Eckart, Peter, 95, 109, 128, 155, 223
Egelhoff, Tom, 23, 167, 181
Eytan, Ted, 13, 39, 70, 140, 176, 188, 246, 261, 277

Faryna, Stan, 60
Featherly, Kevin, 192, 220, 234, 264, 281
Finberg, Howard, 9
Fineman, Ben, 282
Firestone, Charles, 12, 242, 259, 275
Fouts, Joshua, 126, 246, 262
Froomkin, Dan, 37, 278
Froomking, Dan, 261
Fuller, Allen, 265
Fulton, B. Keith, 23, 44, 168, 174, 205, 224

Garfinkel, Simson L., 2, 38, 54, 73, 87, 242, 244, 260
Geith, Christine, 21, 42, 89, 123, 153, 190, 244, 277
Gillmor, Dan, 2, 20, 262, 277
Glaser, Mark, 6, 86, 194, 207, 247, 264, 281
Goodman, Joshua, 10, 18
Gunn, Moira, 9, 17, 105, 120, 151, 205, 243

Halavais, Alex, 43, 89, 152, 167, 175, 192, 221, 233, 249, 262, 279
Halder, Bornali, 61, 74, 109, 142, 167, 179, 195
Hansen, Timothy, 87, 278
Hapgood, Fred, 15, 54, 164, 190, 244, 259, 276
Hargittai, Eszter, 71, 88, 278
Harrison, Candi, 154
Hassencahl, Fran, 43, 246
Hewitt, Perry, 90
Hodhson, Brenda, 261
Hofheinz, Albrecht, 59, 167, 181, 285

Jackson, Nigel, 108, 124, 178,
Jarboe, Ken, 41, 84
Jaroslovsky, Rich, 243
Jenkins, Henry, 126, 153, 166, 178, 206, 233
Joloin, Jerome, 61
Jung, Joo-Young, 89, 222

Kantrovich, Lyle, 8, 44, 74, 110, 168, 180
Kaplan, Daniel, 25, 75, 109
Keith, Kim, 94
Kelly, Mike, 18, 206, 243, 260, 276
Kenyon, Susan, 92, 143
Kreps, Gary, 42, 72, 88, 107, 122, 178, 189, 206, 221, 245, 261

LaFond, Gerald, 250
Lessman, Robert, 167
Levin, Douglas, 9, 88, 124, 248, 265
Levine, Peter, 39, 73, 85, 140, 219, 259, 263, 280
Lieb, Rebecca, 167, 181
Lovelace, Graham, 44, 178, 265, 284
Lunn, Robert, 40, 55, 71, 87, 106, 123, 142, 165, 222, 234, 280

MacLeod, Alec, 93, 224
Mahaffie, John, 45, 179, 194, 285
Maker, Meg Houston, 90, 264
Manafy, Michelle, 7, 89, 125, 222
Mangalmurti, Vikram, 220
Marcus, J. Scott, 107, 138, 205, 232, 260
Markham, A., 151
Marshall, Jon, 225
Metcalfe, Bob, 2, 20, 57, 70, 121, 167

Contributor Index

Mettee, Cory, 128
Miller, Ezra, 109, 127, 166, 177, 194, 246, 282
Mogensen, Kirsten, 263
Moore, Scott, 60, 89, 127
Morgan, Arlene, 178, 245, 264
Murphys, Pat, 234

Ness, Dan, 14, 58, 74, 109, 155, 167, 208, 224, 235, 265, 283
Neubert, Michael, 46

O'Brien, Mike, 10, 143, 181, 209, 224, 265, 274
Opel, Andy, 19, 57, 259
Osterby, Aaron, 195
Otte, George, 108, 125, 220, 233, 245, 264

Peizer, Jonathan, 86, 107, 125, 140, 165, 193, 208, 219, 233, 244
Pena, Carlos Andres, 126
Pickett, William B., 23, 108, 246, 264, 282
Pittman, Terry, 7, 108, 193, 220, 233, 245, 261, 278
Pouzin, Louis, 38, 166, 262
Punnett, Sam, 24, 181, 209, 235, 282

Rainie, Lee, 1, 301
Reed, Mack, 14, 39, 72, 87, 106, 140, 177, 191, 207, 245, 261, 280
Reich, Brian, 11, 41, 72, 86, 106, 122, 141, 153, 177, 206, 221, 245, 263
Reina, Jorge, 38, 58, 139, 174, 218
Rheingold, Howard, 2, 22, 108, 192, 219, 243, 262, 278

Rivero, Victor, 23, 42, 57
Rose, Alexander, 243, 275
Ross, Lorraine, 42
Rovner, Mark, 41, 84
Rushkoff, Douglas, 20, 38, 73, 87, 105, 138, 151, 218, 243, 260
Russell, Cheryl, 13
Rykert, Liz, 152, 177, 248, 262, 279

Samuel, Alexandra, 56, 72, 141, 175, 191, 249, 262, 278
Samuels, Cynthia, 13, 221, 247
Sands, Daniel Z., 12, 260
Sartori, Joao, 110
Schaffer, Jan, 7, 87, 151, 247, 263
Schement, Jorge Reina, 38, 58, 139, 174, 218
Schlieman, Oren, 223
Schroeder, Ray, 155
Schur, Stephen, 91
Scott, David, M., 179, 220, 263
Scott, Travers, 94
Seip, Roger, 283
Shane, Peter M., 11, 22, 38, 260
Shlain, Tiffany, 9, 90, 150, 168, 265, 284
Slavin, Tim, 155, 223
Smith, Barbara, 45, 90
Staton, Elizabeth W., 194
Stewart, William, 16, 59, 75, 110, 127, 143, 168, 179, 209, 250, 284
Straumanis, Andris, 75
Strause, Gordon, 88, 177, 219
Streeter, Tom, 181, 207, 222, 265

Taglang, Kevin, 247
Tarantino, Taryn, 60
Tedesco, Donna, 280
Tewksbury, David, 152, 207, 262
Tracy, Elle, 73, 142, 209

Van Ness, Peter W., 43, 143, 154, 180, 193, 265, 284
Verharen, Egon, 194, 208, 234, 283
Vines, Rose, 21, 73, 85, 124, 176, 192, 208, 279
Virgo, Philip, 25, 56, 126, 138, 166, 207, 219, 246, 263

Warren, Bill, 18
Weinberger, David, 2, 21, 123, 139, 260, 276

Weisman, Mike, 8, 44, 59, 75, 94, 168, 180
Weiss, Daniel, 46, 92, 168, 180
Wellman, Barry, 18, 39, 95, 122, 151, 176, 191, 243, 260
Whitten, Pamela, 188
Witherspoon, Mike, 93, 154
Witt, Leonard, 93, 224, 249
Wollowski, Michael, 71, 166, 282

SUBJECT INDEX

9/11. *See* September 11, 2001, terrorist attacks

access to information, 9, 13, 18, 72, 165, 168
adoption, 28, 124, 158, 188, 204, 231–232, 235, 245, 258, 260–262, 267, 273, 290
advertiser, 57, 245
Age of Creativity, 217–218, 222, 225, 228
Al-Qaeda, 24, 59, 65
"always-on" Internet, 20, 23, 159, 292
American Democratic Project
anonymity, 52, 72, 77, 104, 140, 164, 244, 295
archiving, 194
artists, 15, 22, 80, 217–218, 222, 226, 229, 280, 288
attack, 53–67, 81, 116, 140, 146, 159, 252
automatic identification device, 103

bandwidth, 26, 207, 234, 252, 282, 293
BlackBerry, 157–158, 160, 265
black market, 105
blackout, 57, 64
Bowling Alone (by R. Putnam), 83. *See also* Robert Putnam
brands, 29, 206
broadband, 7, 15, 18, 57, 63, 108, 135, 149, 166, 191, 206, 212–214, 218, 220, 231–234, 236–239, 245–247, 249, 253, 255, 257, 261, 271, 275, 277–278, 290–291

broadband (*continued*)
adoption, 28, 124, 158, 188, 204, 231–232, 235, 245, 258, 260–262, 267, 273, 290
Bush, George W. (President), 148

centralization, 8, 24, 28, 61, 76, 256, 277
chat rooms 6, 49
CIA, 35, 54, 68, 111
citizen media, 6
civic engagement, 83, 87–91, 94, 98, 100
civic life, 87, 98, 241
civic responsibility, 87
civil liberties, 106, 110–111, 113–114, 165, 289
Clarke, Richard, 64
collaboration, 42, 219–222, 234, 244, 251, 277
college students, 122, 124
commercial services, 38
commercialization, 273, 283, 288
communities, 5, 16, 18–19, 23, 37–38, 45, 51, 65, 84, 91, 93, 95, 97, 99, 122–123, 128, 139, 164–165, 168–170, 193, 222, 248–249, 261, 278, 280–281
community groups, 84
computer training, 232
computer virus, 53, 61
Congress, 11, 46, 66, 80, 109, 115, 190–191, 223, 249, 252, 280

connectivity, 7, 21, 23, 88, 107, 150, 158, 205–206, 233, 238, 282, 290–291
consensus building, 180
constitutional, 106, 109, 115
copyright, 24, 69–70, 72, 74–76, 79–81, 222–223, 225–227, 267, 282, 285, 288, 296
cost of access, 232, 239
counterterrorism, 112
creativity, 23, 178, 218–219, 221–223, 225–229, 242, 250, 288, 293. *See also* Age of Creativity
crime, 28, 41, 51, 59, 67, 105–106, 110–111, 113, 115–116, 266
customer service, 20, 187, 189, 193–194, 196–200, 244
cybernetics, 29

datacasting, 204
daycare, 155
decentralization, 8, 24, 28, 76
democratic, 22, 103, 107, 109–110, 112, 114–116, 138–139, 143–144, 167, 171, 181, 279
democratization, 14, 108
Democrats, 145, 175
deployment, 3, 42, 138, 235, 287, 290
developing nations, 16
dial-up, 231–232, 236–237, 257, 291
digital, 6, 8, 11, 14–15, 21, 26, 46, 49, 69–81, 86, 100, 102, 125, 138, 140, 144–145, 150, 159–160, 166–167, 171, 189–190, 200–201, 204, 206–207, 211, 214, 217, 219–221, 227–229, 232, 238–239, 241, 243, 245, 247, 255, 257–258, 261, 276, 278, 286, 290

digital (*continued*)
books, 15, 76, 122, 183, 204, 206, 218, 247, 250, 256, 259, 265, 270, 272
libraries, 133, 217, 220, 225, 227–228, 255, 263, 273, 281
literacy, 111, 127, 188, 193, 218, 235, 248, 278
media titans, 6
music, 5, 15, 19, 69–70, 72, 76–78, 81, 204, 214, 217, 220, 227–228, 242, 261, 265–267, 284–285, 290
news, 2–3, 5–8, 14–16, 26–29, 32, 76, 89, 101, 116, 125, 162, 174–175, 177–178, 183, 185, 186, 192, 204, 210–211, 220, 222, 228, 234, 244, 245, 247, 259, 261, 264, 269–270, 272, 277, 279–281, 288, 291
photography, 15, 218, 286
products, 3, 7, 25, 29, 35, 43, 47, 52, 69–72, 74–78, 80–81, 107, 193, 215, 226, 229, 247–248, 258, 275, 290
prohibition, 70, 76–77
television, 15, 84, 97–98, 120, 155, 181, 183–185, 203–205, 207–213, 215, 218, 220, 238, 247, 252, 264, 268, 271, 283, 292
video, 13, 15, 20–21, 88, 113, 125, 135–136, 151, 160, 190, 192, 203, 205, 207, 211, 214, 225, 235, 241, 250, 252, 255, 264, 274, 277, 283, 286–288, 290–293
digital divide, 14, 26, 102, 125, 138, 144–145, 207, 232, 238, 241, 245, 247, 257, 276

Subject Index

digital rights management, 70, 73–74, 78, 204
distance education, 121–122, 124
divorce, 154
doctors, 190–191, 193–194, 200–202, 271
downloading music, 69
drama, 5, 11, 18, 38, 93, 111, 129, 147–148, 150, 156, 159, 186, 193, 220, 231, 233, 236, 238, 262, 270
DVD, 31, 161, 206, 208, 214, 292

e-commerce, 37, 74, 107, 124, 147, 152, 219, 242, 249, 255, 258, 260, 267, 270, 276
e-mail, 2, 13, 17–18, 25, 28, 33, 39, 43–44, 84, 97, 102, 144, 155–158, 160, 181, 185, 190, 214, 220, 245, 249–250, 252, 258–261, 262, 265–270, 272, 274, 285–287, 292
economics, 10, 136, 140, 242
education, 5, 9, 11, 19, 21–23, 26–28, 30, 42, 57, 80, 91, 120–134, 136, 142, 154, 188, 194, 208, 224, 234, 236, 241–242, 244, 247–248, 250, 254, 256, 267, 283, 285–287, 296. *See also* U.S. Department of Education
education divide
elections, 96, 144, 146, 148, 292
electronic communications, 88
electronic medical treatment, 192
electronic voting, 88, 143
elementary school, 120
embedded computing devices, 108, 115
embedded networks, 105, 108, 110
extremist groups, 169–170

face-to-face communications, 265
family life, 17–18, 153–155
FBI, 111
FCC, 234, 257
file sharing, 56, 70, 72, 79–81, 258, 261, 266, 270, 272
film, 5, 15, 73, 229, 268, 286
financial, 13, 21, 48, 61, 63, 229, 268, 270, 273
formal education, 9, 124, 127, 132
free market, 66
Friendster, 37, 42, 96

geography, 37, 49, 101, 164, 276
globalization, 260
Google, 2, 6, 22, 61, 179, 249, 261, 264, 271–273, 293
government, 1–2, 5, 11, 14, 18, 21–22, 30, 32, 55–56, 58–59, 63, 72, 75, 80, 89, 103–105, 107–109, 111–116, 124, 129, 136, 139, 148, 153, 161, 165, 167, 170, 195, 197, 225, 232–234, 237, 239, 244, 248–250, 255, 261, 266, 273, 283, 288, 291
government subsidy, 232
GPS, 106, 259, 277
graduate school, 120
grassroots, 13, 28, 98, 222
groups, 8, 13, 16–18, 26, 28, 44–45, 48–49, 52, 59, 63, 72, 75, 83–86, 89, 91–94, 96–102, 113, 116, 126–127, 132, 135, 140–141, 145, 150, 163–171, 175, 179–180, 193, 219, 226, 235, 255, 257, 276, 287
guilds, 85, 164–165

health, 5, 12–13, 22, 27, 30, 56, 95, 152, 159, 161, 180, 187–202, 241, 249–250, 252–254, 271,

health (*continued*)
    275, 277, 281–282, 287,
    289–292, 296
health care, 5, 12–13, 22, 30, 187,
    189–192, 194–202, 241, 250,
    252–254, 275, 277, 281, 287,
    291–292, 296
health systems, 12
high-speed connections, 233, 235, 237
high-speed networks, 81, 231,
    234–235, 238
HIPAA, 188
HMO, 193, 197
Hollywood, 75, 229
homeschooling, 149, 151, 153,
    158, 160–161
household, 18, 129, 203–204, 208,
    212–213, 238–239, 243, 260, 282

ideology, 38
individualism, 18
Information Age, 29, 66
information filtering
infringement, 69, 81, 109–110
instant messaging, 13, 28, 185, 259,
    272
institutions, 5, 8–9, 11, 13, 17–19,
    21–22, 25–28, 36, 58, 91,
    120–121, 125, 131, 133, 152,
    154, 185, 190, 197, 200, 244,
    250–252, 281–282
insurance, 188, 190, 192–196,
    198–200, 281
intellectual property, 29, 37, 73–76,
    79–80, 107, 124, 152, 219,
    246, 262, 267, 270, 275–276,
    285, 288
international relations, 5, 16
Internet access, 119–120, 144, 155,
    238, 253, 257, 260, 275, 287
Internet connections, 156, 257

Internet marketing, 20, 60
Internet service provider, 56
Internet solicitation, 98
iPod, 15, 215
Islam, 62
ISP. *See* Internet service provider
iTunes, 73, 78, 284

journalism, 1, 6, 8, 248, 263–264

K–12 education, 124, 129, 247,
    250
Kazaa, 15

learning, 9, 12, 45, 119–136, 181,
    188, 201, 241–242, 249–251,
    256, 282, 284, 287
leisure, 33, 87, 100, 149–161, 205,
    264, 271, 296
licensing, 12, 76, 81, 288
literature, 1, 5, 217–218, 224, 228,
    279, 281
litigation, 79, 196
local government, 59

magazines, 210, 270
mainstream media, 181, 263
malpractice, 187, 190–191, 193, 197
malware, 57
McCracken, Grant, 77
media, 6–8, 11–15, 17, 19–22, 24,
    27–28, 30, 32–33, 39, 44, 64, 73,
    78–79, 81, 87, 91, 93, 95, 116,
    128, 135–136, 150, 155, 166–167,
    174, 176, 178–185, 203–215,
    218–220, 222–223, 225, 236,
    238, 242, 245, 247, 250, 252,
    254, 256, 258, 260–261,
    263–265, 267, 269, 271–272,
    279, 281–285, 288, 293
media convergence, 203, 206

Subject Index

media toys, 203
medical mistakes, 187, 194, 199
medicine, 5, 12, 27, 31, 188, 191–195, 199, 201, 252–253, 275, 282, 290, 293
message boards, 6
micropayments, 77, 81
Microsoft, 2, 10, 18, 61, 67, 72, 258, 285, 292
Middle East, 14, 283
military, 5, 17, 22, 27, 55, 62, 114, 259
millennial generation, 93
mobile devices, 31, 49, 150, 269
monitoring devices, 112
MP3, 208, 292
MPAA, 75
music, 5, 15, 19, 69–70, 72, 76–78, 81, 204, 214, 217, 220, 227–228, 242, 261, 265–267, 284–285, 290

Napster, 7, 15, 74
narrowcasting, 169
National Security Agency, 365
neighborhoods, 5, 276–277
network infrastructure, 65, 67, 170, 250
network security, 137–139, 141–145, 147–148, 287
networking, 28, 38, 45, 77, 81, 88, 93, 99, 113, 166, 175, 177, 212, 219, 260, 262, 292
news organizations, 5–6, 247
newspapers, 6, 14, 175, 177, 183, 185, 204, 261, 270, 277
NGO, 87
NSA, 106

offshore tech support, 16
online education, 122, 256
online gaming, 6, 16

online medical resources, 187, 194
online security, 36
online universities, 131
open development, 20
open-source access, 72

partisan press, 174
Patriot Act, 106–108, 111, 115
PC, 27, 49, 73, 85, 124, 132, 176, 192, 204, 208, 210, 213, 219, 232, 258, 262, 265, 279, 287
peer support, 188–189, 199
peer-to-peer networks, 69, 71, 73–74, 80–81, 266
personal entertainment, 205, 209, 213
personal interaction, 38
physicians, 12, 193, 255
piracy, 55, 73, 75, 80, 195
podcasting, 80
polarization, 174–176, 181–182, 243, 271, 285
political deliberation, 173
politics, 5, 11, 13–14, 19, 22, 24, 26–27, 38, 90, 95, 98, 138, 140, 143, 163–165, 178–179, 183, 243, 248, 257, 260, 291
portable media devices, 210
poverty, 14, 119, 236, 286
power grid, 53–56, 58–63, 66–67, 214
prescription, 190–191, 194
presidential campaign, 87, 99
privacy, 17, 21, 23, 36, 42, 46, 50, 58, 76–77, 108–117, 144, 193, 195, 198, 244, 247, 275, 278, 295
private property, 70
productivity, 19, 152, 154, 161, 265–266, 273
public networks, 48, 204
public opinion, 27, 127, 261

public schools, 119
publishing, 5–9, 11, 14, 28, 72, 177, 229, 258, 261, 286–287
Putnam, Robert, 83, 92, 95

radio, 9, 15, 84, 94, 97, 105, 120, 183, 185, 199, 207, 212–215, 242, 244, 283
radio frequency identification (RFID), 105–109, 114
really simple syndication (RSS), 28
recording industry association, 69
regulation, 42, 64, 108, 115, 198, 251, 292
religion, 5, 19, 38, 101, 163–164, 167
Republicans, 145, 148, 175
research, 1–3, 11, 14, 17, 25, 28, 36–37, 40, 42, 45, 55, 70–71, 87, 90–92, 102, 106, 122–123, 128, 130, 136, 142, 147, 165, 188, 193–194, 199, 208, 222, 234, 248, 256–257, 262, 273, 280, 283, 289
RFID. *See* radio frequency identification
rising health care costs, 187, 189, 200
RSS. *See* really simple syndication

Schneier, Bruce, 58
school, 9, 18, 20, 41, 70, 96, 119–120, 122–129, 131–136, 149, 151, 153–156, 158–161, 236, 267, 276, 282
science, 26, 36, 54, 70, 164, 222, 224, 228, 276, 282
search engine, 266, 289, 291
self-publishing, 221
September 11, 2001, terrorist attacks, (9/11), 54, 64, 94
single parents, 154
social activities, 41

social interaction, 20, 36, 40, 48, 95, 99, 135, 157
social mobilization, 95
social networks, 3, 8, 35–42, 44–47, 49–52, 93, 96–97, 246
software, 56–57, 61–62, 72, 74, 76, 79, 80–81, 130, 142–143, 192, 210, 229, 247, 261, 265, 268, 273–274, 276, 285, 291
spam, 26–27, 36, 73, 79, 184, 241, 244, 251–253, 258–259, 261, 263, 266, 269, 278, 280, 285, 287, 290, 292
special-interest groups, 13, 98
spyware, 57, 81, 113, 116, 244
student-directed learning, 132
students, 1, 2, 11, 41, 91, 119, 120, 122–124, 126–130, 132–136, 150, 155, 246, 264, 273, 288
Supreme Court, 106
surveillance, 17, 23, 103–116, 140, 165, 244, 278

teaching online, 120
Tech Nation, 9, 17, 105, 120, 151, 205, 243
technology, 1–3, 7, 8, 10, 14, 20–21, 23–24, 28, 30–31, 40, 42, 45–46, 54, 57, 60–61, 65, 71, 73, 75, 77, 85, 90–92, 94, 104–106, 108–117, 120, 124–125, 127–131, 133–135, 138–142, 148, 150, 152, 155–159, 163–164, 169, 171, 176, 178, 180, 182–183, 186, 188, 190–192, 194–198, 200, 204, 207–208, 212, 214, 220, 222–223, 226–229, 233–234, 236–238, 243–245, 247, 251–252, 255–256, 259, 262, 264–266, 268, 271, 275–277, 279–280, 283–286, 288, 291–293, 296

Subject Index

telecommuting, 10, 149, 155
telemedicine, 194–195, 253, 275
television, 15, 84, 97–98, 120, 155, 181, 183–185, 203–205, 207–213, 215, 218, 220, 238, 247, 252, 264, 268, 271, 283, 292
terrorists, 54, 56–57, 62–64, 66–67, 153, 289, 295. *See also* September 11, 2001, terrorist attacks
textbooks, 247
time crunch, 84
TiVo 15, 204, 210, 276
transportation, 249
Trojan horse, 54
trust, 3, 23, 35–43, 45–52, 58, 65, 71, 136, 139, 143–145, 147–148, 159, 168–170, 174, 193, 251, 277, 283, 295
TV. *See* television

U.S. Department of Education (DOE), 131

vacation, 50, 157, 160
video games, 88, 125, 203, 287
videoconferencing, 13, 151, 190, 205
virtual classroom, 120, 125, 135
virtual textbook, 150
viruses, 53–55, 61, 241, 244, 253, 258–259, 263, 266, 287
voice over Internet protocol (VoIP), 21
VoIP. *See* voice over Internet protocol
volunteerism, 86
vulnerability to attack, 53, 62, 64–65, 140

Weblogs, 6, 14, 264, 266
Wi-Fi, 49, 233, 257, 259, 274, 278, 292
wireless, 21–22, 25, 27–28, 110, 160, 204, 207, 209, 233–234, 237, 239, 262, 264–265, 282–283, 286–287, 292
wireless networks, 233, 239, 283
women, 14, 150, 215, 264
workplaces, 5, 32, 122

Y2K, 57, 246

zealots, 109, 163–164, 166, 168, 170, 182

LaVergne, TN USA
19 January 2010
170412LV00003B/29/P